QUALITY AND COMPETITION

LAWRENCE ABBOTT

QUALITY AND COMPETITION

An Essay in Economic Theory

New York 1955

COLUMBIA UNIVERSITY PRESS

FIRST PUBLISHED IN BOOK FORM 1955

Library of Congress Catalog Card Number: 55-7945

PUBLISHED IN GREAT BRITAIN, CANADA, INDIA, AND PAKISTAN
BY GEOFFREY CUMBERLEGE: OXFORD UNIVERSITY PRESS
LONDON, TORONTO, BOMBAY, AND KARACHI

MANUFACTURED IN THE UNITED STATES OF AMERICA

Preface

THIS book is the result of thought and writing that span a period of about twelve years, and that have undergone a good many transformations. The individual ideas in it, generally speaking, are not new, although it happens that a number of them were worked out independently and only subsequently were found to have been held by others. Whatever is novel and useful in this work lies in the grouping and elaboration of the ideas. As a collectivity they provide a pathway to conclusions foreign to, and in some cases opposite to, the generally accepted conclusions of contemporary economic theory. More important, they provide tools for the study of certain economic problems and bench marks for the making of welfare judgments which, it is hoped, will be considered superior to those now possessed by the economic theorist.

I am deeply conscious of the debt I owe to others, and of the impossibility of acknowledging all the sources of help on which I have drawn. I am under special obligation to John Maurice Clark, whose guidance and critical insights have assisted me over many rough spots, to George J. Stigler and Donald W. O'Connell for numerous helpful criticisms, and to Horace Taylor for his early encouragement which made possible the development of these ideas. Responsibility for the present form and content of the book, and for any errors or shortcomings in it, is of course my own.

LAWRENCE ABBOTT

Union College
Schenectady, New York
January, 1955

Contents

QUALITY AND COMPETITION

The deductive science of Economics must be verified and rendered useful by the purely empirical science of Statistics. Theory must be invested with the reality and life of fact. . . . But, before we attempt any investigation of facts, we must have correct theoretical notions.

—WILLIAM STANLEY JEVONS
The Theory of Political Economy

Introduction

THE basic ideas in this book are quite simple. But they are the sort of ideas that may not be immediately acceptable to most economists, for they collide sufficiently with orthodox economic thinking to require of the reader some rather fundamental—and perhaps painful—readjustments in his approach to the study of economic behavior and market structures and their welfare implications. For this reason the book cannot be as brief as the writer would like to make it, nor as easy reading.

The impetus which has brought it into being has been the writer's conviction that a serious gap exists in the structure of economic theory. For almost as many decades as economics has existed, economists have been so preoccupied with questions of price determination and price-output relationships that the equally interesting and important phenomena relating to the "product" or "quality" variable have received but scant attention. The result has been that economists now have at their disposal a well-integrated body of analysis dealing with price behavior in various types of markets, but have no comparable body of analysis to guide them when studying product behavior.

This shortcoming cannot be lightly dismissed. A complete understanding of competitive markets involves an understanding of both the price-adjustment aspect and the product-adjustment aspect, and requires a framework of analysis comprehensive enough to embrace both aspects. Until such a framework is built, the economist who is concerned with the extent and character of competitive forces in capitalist markets must hobble along with only a partial theory in his analytical toolbox—a "theory of competitive price"—instead of a complete, well-rounded "theory of competition."

The present book represents a modest attempt to work toward the formulation of a more comprehensive theory. Its main concern is the neglected half of the study of competitive markets: analysis of the quality variable. Its primary objective will be to work out a theory of quality determination under competitive conditions. Only to a minor extent will it be concerned with the effects of quality differences and changes on price. And only in the final portion of the book will an attempt be made to employ quality analysis in combination with price analysis.

The term "quality" will be used in this book in its broadest sense, to include all the qualitative elements in the competitive exchange process—materials, design, services provided, location, and so forth. Quality analysis embraces both quality differences between competing products (commonly termed "product differentiation," "product heterogeneity," or the existence of "imperfect substitutes") and temporal qualitative changes in products ("product variation" or the creation of "new products" or "new models"). Both of these elements enter into what is known as "quality competition," which in turn is one of the two elements in "nonprice competition," the other being advertising and sales promotion. With this second aspect of nonprice competition we shall also be concerned, since it is intimately bound up with quality comparison and quality competition; the chief emphasis, however, will be on the products themselves.

This is an essay in theory, not an empirical study of nonprice competition. It is concerned not so much with the facts of the world as with ways of thinking about them. Theory, of course, is only a first step. It can do no more than devise hypotheses; only by resort to facts can the usefulness of hypotheses be determined. It seems best, however, to confine this study to questions of theory, partly in order that its size be kept within bounds, and partly because of the complexity of the problem of verification. As the reader will himself discover in due course, it is no easy task to determine the kind of empirical data needed in order to test the propositions developed in these pages.

In two other ways the scope of this study will be restricted. It will be concerned almost exclusively with the economics of a competitive capitalist society. At only one or two points will the economics of a socialist economy be considered. And it will analyze the "efficiency" of capitalist markets of various sorts from one specific standpoint: that of the public welfare. The reader will be asked to judge the usefulness and adequacy of the analysis solely on the basis of its applicability to practical questions of economic welfare and public policy.

Even so, this is a large order. It would be presumptuous to claim that this book provides definitive answers to the questions raised. Its purpose may be regarded as achieved if it does no more than reveal gaps in the existing body of theory and indicate the direction in which further study must proceed in order to fill the gaps.

Part One

QUALITY VARIABILITY AND EXISTING ECONOMIC THEORY

Chapter I

THE INADEQUACY OF THE PRESENT COMPETITIVE IDEAL

1. PURE OR PERFECT COMPETITION AS A STANDARD

THE core of theoretical economics has been, from the time of Ricardo and Cournot up to the present Keynesian age, the theory of competitive markets; and the kind of competition envisaged in this theory has been the kind that occurs in the great produce markets, where many buyers and sellers dicker over the terms of sale of quantities of some homogeneous or standardized commodity whose quality is never called into question.

One rather striking feature about such markets is the complete absence—at least from the picture as presented in economic analysis—of quality differences, quality variations, and hence of quality comparisons. Presumably the buyers know what grade they are buying, are aware of the existence of other grades and of other commodities that might serve the same general purpose, and have already made their selection on some rational basis. Yet these matters, according to the usual economic analysis, are not part of the competitive market process, but are to be explained by the principle of Substitution. In the market itself we find only Competition in its pure, unalloyed form. And since the competing products, no matter from what source they come, are considered to be identical, the only variables that are subject to determination in the market are price and output. Thus the conditions which define such a market also define the single kind of competition that can occur in it: price competition. This kind of competition has been enthroned in the theoretical literature of economics as "pure competition" and "perfect competition," and has been held up by welfare economists as the paragon of competition.

This enthronement, however, can be—and has been—seriously questioned. The first and most obvious criticism to be made is that, even if such markets do embody competition in its purest form, they are too rare to be regarded as representative of the kind of competitive exchange process that predominates in a real-world free-market economy. If we turn our gaze from staple commodities to fabricated producers' goods and to most kinds of consumers' goods, we usually find qualitative differences between the rival products being offered on any given market, giving rise to phenomena not to be found in the great wholesale commodity markets: investigation and quality comparison on the part of buyers, sales promotion and advertising on the part of sellers, and also, from time to time, qualitative alterations in the products themselves. All this is too well known to require elaboration. Equally well known is the growing realization by economists during the second and third decades of the present century that economic theory in this respect was seriously out of step with reality. As far back as 1918 J. M. Clark pointed out that "there is so wide a field in which a difference between the goods offered by the different makers is one of the essential features of the competitive struggle that this is really the typical case rather than the exception." [1] In the early 1930's a frontal attack on the then-existing *Weltanschauung* of economic theory was launched with Joan Robinson's *The Economics of Imperfect Competition* and Edward H. Chamberlin's *The Theory of Monopolistic Competition.* The revolution was swift. It soon became widely held that "monopolistic conditions are to be regarded as ubiquitous and determining, rather than exceptional and inconsequential, features of modern economic life." [2] The net result has been to reject both pure competition and complete monopoly as models of reality. The inter-

[1] John Maurice Clark, "Economics and Modern Psychology," *Journal of Political Economy,* 1918, reprinted in *Preface to Social Economics,* p. 110.

[2] Horace G. White, Jr., "A Review of Monopolistic and Imperfect Competition Theories," *American Economic Review,* XXVI (Dec., 1936), 637.

mediate worlds of monopolistic competition, pure oligopoly, and differentiated oligopoly now receive the major share of attention.

Yet, in the bulk of contemporary economic literature, the result has not been to dethrone pure competition, but simply to transfer its position from that of the *typical* competitive situation to that of the *ideal* competitive situation. Much of recent monopolistic competition literature either explicitly or implicitly adopts pure competition as a standard of comparison which is considered to be appropriate and at least conceptually applicable to monopolistically competitive situations. Welfare economics, too, has accepted perfect competition as a standard of what is desirable and has buttressed its conclusions with mathematical demonstration.

Such acceptance of pure or perfect competition as a norm has, of course, obvious implications. Market situations of the real world which deviate from pure competition must be considered inferior. Thus "pure competition"—frequently referred to, simply, as "competition"—becomes the basis for value judgments about specific competitive situations and economic institutions. Judgments resting on this foundation are commonplace in the economic literature of recent years. As one welfare economist recently expressed it, "The theory of Monopolistic Competition, as it is usually developed, shows (implicitly) how an energetic monopolist can diminish the community's welfare." [3]

What the theory actually shows is that, even aside from situations having oligopolistic elements, whenever the competing products in a market (defined in some sense) are not homogeneous, individual sellers have negatively sloping demand curves, with the result that (on the assumption of U-shaped cost curves) "the price is inevitably higher and the scale of production inevitably smaller under monopolistic competition than under pure competition." [4] This proposition is frequently interpreted to mean that freedom of entry, mobility of resources, and competitive pressures combine to drive each producing firm to a situation in which resources are not

[3] Melvin W. Reder, *Studies in the Theory of Welfare Economics*, p. 88.

[4] Edward H. Chamberlin, *The Theory of Monopolistic Competition*, 6th ed. (hereinafter cited as Chamberlin, *Monopolistic Competition*), p. 88. See also pp. 114, 166, 175–76.

being used to the best advantage, and therefore that product and service differentiation is wasteful from the social point of view.[5]

How serious are the deviations from the optimum? Are they large enough to have implications for public policy? Rather widespread has been the answer in the affirmative. Oskar Lange, arguing the case for socialism, goes so far as to declare that "the actual capitalist system," being "one where oligopoly and monopolistic competition prevail," is "far removed from the model of a competitive economy as elaborated by economic theory" and "will appear to a future generation as the craze and folly of a past epoch." [6] Even among those who take a more moderate view of the effects of monopolistic competition, conclusions drawn from monopolistic competition analysis have been rather widely accepted as furnishing a guide for public policy.

A further divergence of the competition typical of the real world from the theoretical norm occurs as a result of expenditures on advertising and sales promotion. Such expenditures are inconsistent with the assumptions of the economist's model of ideal competition. Advertising is thus associated with the monopolistic element in capitalistic markets. Sometimes it is referred to as a deliberate device for differentiating products which might otherwise be regarded as indistinguishable. Sometimes it is viewed as a sales-capturing device made possible by already-recognized differences between competing products—a situation which furnishes an opportunity for the enhancement of monopoly control not to be found under pure competition. But whatever the motive for advertising expenditures, the economic effect—aside from any possible interference

[5] Chamberlin has mentioned several specific respects in which, he claims, monopolistic competition has undesirable consequences: for instance, "the impossibility of production under the most efficient conditions" which "is settled once and for all by the shape of the demand curve" (*Monopolistic Competition*, p. 98); the fact that "for a given price, 'product' is inevitably somewhat inferior" (*ibid.*, p. 99); and the existence, under certain circumstances, of excess productive capacity, which leads him to comment that "wastes in the economic system . . . which are usually referred to as 'wastes of competition' . . . could never occur under pure competition. . . . They are wastes . . . of the monopoly elements in monopolistic competition." (*Ibid.*, p. 109). He has not claimed, however, as others have, that monopolistic competition is *on balance* less desirable than pure competition. With reference to this point, see below, p. 13.

[6] Oskar Lange, "On the Economic Theory of Socialism," in Oskar Lange and Fred M. Taylor, *On the Economic Theory of Socialism*, pp. 107, 108.

with the degree of "purity" of competition—is clearly a cost which has no counterpart in pure competition.

2. SOME OBJECTIONS TO THE STANDARD

Yet the claim that pure or perfect competition is the most desirable form of competition has not gone unchallenged. According to some authorities the *absence* of differentiation, variation, and promotion of the competing products may be a sign of shortcomings in the market situation. "In so far as consumers really want the variety of product," Donald H. Wallace has written, ". . . the ideal competitive equilibria of traditional theory would not adequately meet their wants. What would appear to be monopolistic exploitation according to the familiar criterion turns out to be a symptom of greater precision in the satisfaction of wants." [7] Chamberlin too has noted that "wherever there is a demand for diversity of product, pure competition turns out to be not the ideal but a departure from it," and has emphasized "the necessity of substituting for the concept of a 'competitive ideal' an ideal involving both monopoly and competition." [8] J. M. Clark, an outspoken critic of monopolistic competition orthodoxy, has likewise argued that some of the so-called imperfections in the theoretical scheme of contemporary economics may turn out to have some socially beneficial effects in actual situations, and that quality competition in particular is a decidedly healthy form of competition. "If every seller," he writes, "is free to choose whether he will offer a quality close to his rivals' or markedly different, if the buyer is equally free to choose between them, and if this rivalry forces every seller to offer something that will attract free buyers, the rivalry is surely one of the most desirable sorts." [9]

[7] Donald H. Wallace, "Monopolistic Competition and Public Policy," *American Economic Review: Papers and Proceedings*, XXVI (March, 1936), 81.

[8] Edward H. Chamberlin, "Product Heterogeneity and Public Policy," *American Economic Review: Papers and Proceedings*, XL (May, 1950), p. 92; *Monopolistic Competition*, p. 214.

[9] John Maurice Clark, *Alternative to Serfdom*, p. 77. This quotation and others from Clark's *Alternative to Serfdom* are reprinted with the permission of Alfred H. Knopf, Inc.

The fact that this type of competition is a spur to economic progress has also been noted. The great champion of this viewpoint was, of course, Joseph A. Schumpeter, whose opinion it was that "in this respect, perfect competition is not only impossible but inferior, and has no title to being set up as a model of ideal efficiency." [10]

With regard to advertising, too, the conclusions derived from value theory have not been wholly acceptable to everyone. If we examine the economic literature not for derivations of theorems but for expressions of opinion, we find that economists generally regard advertising as being sometimes, though by no means always, beneficial. Here again, what theory points to as an imperfection, associated with deviations from the welfare optimum, may also be regarded as an institution which performs a real service whose value is perceptible upon appeal to common sense yet somehow fails to get included in the accounting records of theoretical economics.

On another score doubts may arise as to whether pure competition is acceptable as a standard of comparison. Allusions to "a perfectly competitive system" or "a world in which there is perfect competition in every way" carry the implication that such a system is conceivable, and that, even if it is not to be found in our admittedly imperfect world, an idealized society can be imagined in which deviations from pure competition are wholly absent. But can it? Even if there were no problems of ignorance and irrationality, or of friction and immobility, or of large-scale production and oligopoly, could a society exist which conformed to perfect competition to the extent that homogeneity of the competing products was a characteristic of every market? And if not, is not "a perfectly competitive system" a meaningless idea, inappropriate as a basis for comparison?

Absolute homogeneity, of course, is impossible. In nature no two things are ever exactly alike. Wheat grown in two different fields will never contain exactly the same mineral, vitamin, and protein content. Handmade products likewise vary in shape, quality of workmanship, etc., even when they are meant to be alike and are produced by skilled craftsmen. Machine-made parts also are only approximately alike, depending on the degree of tolerance permitted. Even if all products were carefully classified into numerous

[10] Joseph A. Schumpeter, *Capitalism, Socialism, and Democracy*, p. 106.

types and grades, differences would still be found within each group. The only kind of "commodity" which is truly homogeneous is the kind whose value lies in the privilege accorded to its owner of claiming some stipulated benefit—for example, postage stamps, railroad tickets, securities. Aside from these exceptional cases, homogeneity is merely a convenient simplification, like the statistician's device of lumping different data into a single class.

From the economic point of view, of course, "virtual" homogeneity is as good as absolute homogeneity; the essential condition is that the difference between rival products be so small that the buyer does not find it worth while to distinguish between them. But even virtual homogeneity seldom exists. The best examples of it are raw materials, which in many cases are homogeneous for all practical purposes. Yet even such a thoroughly inspected and standardized commodity as milk can vary considerably in quality. Besides, in any highly organized society such products account for only a minor fraction of the total number of exchanges. In most markets, whether for producer or consumer goods, the products offered have been processed or fabricated or are compounded of several ingredients or are (or include) services—for instance, bread, shirts, machine tools, paint, typewriter repair. Here neither exact nor approximate homogeneity of product is to be found, even among competing goods of the same type, size, and grade.

Furthermore, it would never come about under active and completely free competition. It could only be achieved by force of authority, such as government standardization or grading. And then its attainment would by no means be certain. Uniform quality is difficult to maintain in manufacture, even within a single firm; it requires constant vigilance on the part of management; to maintain it on an industry-wide scale would be far more difficult. The possibility of defects and the fact that goods on display frequently become shopworn are factors that make quality comparison on the part of buyers desirable. Unless grading is carried to absurd lengths, some leeway is left for differentiation within the limits of a single grade.

More important still, there are a good many kinds of products—books, toys, magazines, women's dresses, restaurant meals, etc.—

in which absence of product variation or quality differences is unthinkable. In the above instances variety and novelty are demanded by the buying public. It would be preposterous, for example, to imagine the toy industry deciding to produce a single "standardized" toy, or even a given number of such toys. It would be equally difficult to think of books being sold under conditions of pure competition. Even the sale of a noncopyright classic by numerous competing publishers would hardly take the form of price competition between perfect substitutes. Aside from the question of the literary accuracy of each competing edition, opportunities arise for quality differences in the books' dimensions, paper (weight, whiteness, texture, watermark pattern, freedom from defects) and typography (size and style of type, width of line, spacing between lines, style of chapter headings)—all of which may add up to significant differences in readability and attractiveness between one publisher's edition and another's. In the sale of theater tickets pure competition is not merely unlikely but inconceivable, since differences in desirability of the various locations cannot be ignored. Probably the most heterogeneous commodity of all is human labor. It can be classified roughly for convenience; but attempts, whether by labor unions or others, to standardize it rigidly cannot obliterate the fact that each worker has his own particular set of assets and liabilities.

The mere existence of imperfect substitutes, of course, is not inconsistent with the conditions of pure or perfect competition. If many sellers offered each of numerous varieties of a product, each variety could be said to be sold under conditions of pure competition. Such a state of affairs might conceivably exist in some sectors of an economy. But in many markets the possibilities for further product variation or the impossibility of standardization (as in the case of physicians or radio entertainment), to say nothing of the diseconomies of small-scale production, would make it a very strange kind of world in which *all* products would be sold in purely competitive markets.

Even in the cases of competing products that are considered to be physically identical—where standardization, say, is workable and efficient—buyers may have good reason to prefer one source of supply to another. Chamberlin has pointed to the pervasiveness of differences between one seller's product and another's "with respect to the conditions surrounding its sale . . . convenience of the seller's location, the . . . character of his establishment, his way of doing business, his reputation for fair dealing, courtesy, efficiency."[11] Sometimes differences of this sort are interpreted as evidences of buyers' irrationality. Yet it need hardly be argued that a firm which rectifies errors in shipments and replaces defective merchandise promptly and cheerfully performs a real, not an illusory, service, and that competing firms are not likely to be identical in such matters. Furthermore, producers are extremely unlikely to be identical in their ability and willingness to maintain uniformity of product. How well they do so depends partly on the quality of engineering skill and supervisory personnel employed and partly on that intangible something which may be called the character of the management. For this reason, even if yesterday's output of the competing firms were found to be homogeneous, buyers could not be certain that today's output would turn out to be homogeneous. As to location, the hardly controvertible statement that "retail shops, for example, could not all be located on the same spot"[12] points to the inconceivability of pure competition in retail markets. In producer-goods markets locational differences may not create differences in convenience, yet they frequently give rise to differences in transport costs. It is probably true that in most cases spatial elements alone are sufficient to prevent pure competition from existing in markets where the products of rival firms would otherwise be regarded as homogeneous. The kind of "ideal world" in which pure competition would be the typical state of affairs apparently must be a world without spatial dimensions!

One may argue of course that while it is impossible to conceive of *attaining* pure competition it is nevertheless quite possible to conceive of *approaching* it, and that this possibility makes pure or perfect competition a meaningful norm. This argument would assuredly have cogency if the forces that make for progress and improved welfare in a society also tended to diminish the extent or magnitude of product heterogeneity.

Actually, however, the reverse seems to be true. Homogeneity of

[11] Chamberlin, *Monopolistic Competition*, p. 56.

[12] *Ibid.*, p. 214.

product is most likely to be approximated in the markets for simple, basic materials (that is, raw materials made roughly homogeneous by grading or processing), and least likely to be found in the markets for fabricated products. Until the twentieth century most markets were of the first sort; fabrication was largely a household occupation. But with the growth of modern industrialism fabrication has become more and more the province of the producer rather than the consumer. The fact that so large a proportion of the products in consumer markets today are complex—and hence afford opportunities for diversity—is due largely to the fact that households today buy bread instead of flour, sweaters instead of wool yarn, ready-made cement blocks instead of cement, etc. Technological progress has also contributed to the increasing complexity of goods. Automobiles, for example, are more complex than carriages; modern central heating is more complex than the old-fashioned parlor stove. Consequently there are more ways in which competing goods can differ from each other than in earlier, simpler times.

Another factor that has contributed to the secular growth of product differentiation has been the development of uniformity of output. So long as the goods of a single producer varied unpredictably in quality and design, buyers had little to gain by differentiating one producer's goods from another's. But with the development of precision methods and the adoption of the practice of segregating seconds, the products of a single manufacturer have become more nearly homogeneous, and the differences between them and the products of rival manufacturers have become more significant.

Still another reason may be found in the fact that as man becomes more civilized he develops greater powers of perception with regard to quality differences. Subjective homogeneity may exist even when objective homogeneity does not, due to the inability or unwillingness of buyers to perceive differences between almost identical products and discriminate between them. In a primitive society the niceties of discrimination have not yet been developed, and so a relatively large area of subjective homogeneity is understandable. But as a society matures and education improves, people learn to develop more acute powers of discrimination. Their wants become more detailed. They begin to pride themselves on being connoisseurs in certain fields—to develop a preference, say, not simply for white wine, but for 1948 Chablis.

It is true that imagined differences may be trivial or even nonexistent and that overzealous selling activity sometimes produces a kind of pseudo-differentiation through which buyers are the victims of illusion. The result can hardly be called progress. Some writers imply that the differentiation in modern consumer-goods markets is largely of this sort. In so far as their conclusions rest on value judgments as to what is or is not trivial, they can neither be supported nor refuted. Chamberlin maintains that ill-informed consumers are as likely to overlook real differences as to imagine differences where none exist.[13] To anyone who has attained technical competence in some field this viewpoint must appear sound. People generally tend to underestimate the significance of apparently trivial differences in fields in which they are not expert. An unmusical person may be unwilling to concede that there is any difference in tone between a Steinway and a Chickering piano, being unable himself to detect it. A nongolfer is more likely than a habitual player to believe that all brands of golf balls are virtually alike. In a society composed wholly of intellectual giants, even if the illusory differentiation created by modern selling techniques were wholly absent, product heterogeneity might well be even greater than it is in our present world, and would certainly still be the rule, not the exception.

Pure competition, then, is an ideal far better suited to a primitive society than to one with a highly developed culture and technology. And progress makes pure competition less appropriate as an ideal with every passing year. In a highly industrialized society, the vast majority of markets are markets in which *by the very nature of things* competing products are not and cannot be homogeneous.

In short, a market-by-market examination of the real world *and of any idealization of it that the mind can grasp* leads to the conclusion that (1) an economic system which is "perfectly competitive" throughout is a nonsensical concept; (2) variety of product is a

[13] Chamberlin, "Product Heterogeneity and Public Policy," *American Economic Review: Papers and Proceedings*, XL (May, 1950), 88.

normal situation in most competitive markets; (3) even when arrangements are made to curb or stifle heterogeneity, there are factors at work inherent in the very nature of economic life which tend to make such arrangements only incompletely successful; and (4) product heterogeneity frequently yields benefits that would be lost if the competing products were made homogeneous. Economic theory must somehow find room for the idea that product differentiation and variation form an integral part and desirable feature of any conceivable kind of exchange economy. The first step is to reject the "perfectly competitive system" as a standard of evaluation.

This is not to deny the great usefulness of the concept of pure competition as a tool of analysis in handling many specific economic problems—especially those dealing wholly with quantitative matters, such as prices, costs, outputs. But to regard the so-called purely competitive market as the "standard" or "normal" kind of market, which serves as an indicator of the desirable direction of change, and the so-called monopolistically competitive market as a deviation or distortion, as if its conditions were artificially contrived or could be eliminated, is to turn our backs on the obvious facts of the modern world.

Chapter II

THE INCOMPLETENESS OF PRICE THEORY

IF pure competition must be abandoned as a yardstick, can theory furnish an acceptable substitute? The literature of monopolistic competition contains no satisfactory answer to this question. In it we find the premise that quality differences are a monopolistic element coupled with the premise that the variety made available by quality differences is desired and desirable. But monopolistic competition theory provides no method by which one can distinguish different kinds or degrees of monopoly as being "good" or "bad," or determine whether a competitive situation contains "insufficient" or "excessive" amounts of monopoly. It has been admitted that "theory affords an answer neither as to how far differentiation will 'naturally' be carried, nor as to how far it should be carried." [1]

The recent tendency has been for economists to turn their backs on formal theory in attempting to set up some practical standard of "workable competition" by which to judge the social efficiency of competitive markets. Yet it would be more satisfactory if a workable standard could be derived from theory.

There is evidently something missing from contemporary economic analysis that needs to be included if economic theory is to provide help in the study of questions regarding welfare, policy, and market structures. In the paragraphs that follow I shall attempt to identify this missing "something" with a branch of economic inquiry which has been almost completely ignored in the theoretical literature.

[1] Chamberlin, *Monopolistic Competition*, p. 273.

1. LACKING: A THEORY OF QUALITY DETERMINATION

Questions about product quality for which answers are needed in order to gain a thorough understanding of economic activity may be grouped into two broad classes. One class consists of questions having to do with the *causes* of product quality. What determines the qualities of the particular set of products being produced in an economy at any given time? Under what conditions does product heterogeneity tend to arise, or diminish? What causes producers to alter quality? What role do buyers play in quality determination? And so forth. The other broad class of questions has to do with the *effects* of product heterogeneity and variation on other economic variables. How do quality differences and changes affect prices, costs, profits, output, price flexibility, concentration?

Questions of the latter sort have been rather exhaustively explored in the past two decades. By comparison, only the most meager attention has been devoted to the other type of question. A few scattered attempts have been made to set up models that treat product quality as a variable.[2] But in most discussions of monopolistically competitive markets the existence of product differentiation is simply taken for granted as part of the contemporary economic scene. The customary procedure is to study the processes and adjustments of a market economy under the twin assumptions that, on the one hand, qualities remain fixed, and, on the other, prices and outputs are freely variable. Under these assumptions a theory of price emerges while a theory of quality is impossible. We are led to infer that quality is determined either by noneconomic forces or by economic forces too obscure to analyze.

[2] Among them: Chamberlin's brief but highly suggestive analysis of "product equilibrium" in *Monopolistic Competition* (pp. 78–80); his "The Product as an Economic Variable," *Quarterly Journal of Economics*, LXVII (Feb., 1953); Hans Brems' *Product Equilibrium under Monopolistic Competition;* Peter O. Steiner's "Program Patterns and Preferences, and the Workability of Competition in Radio Broadcasting," *Quarterly Journal of Economics*, LXVI (May, 1952); the analyses of spatial competition by Hotelling, Smithies, and others.

2. SOME REASONS FOR THE NEGLECT

To anyone who is familiar with the development of economic thought, this neglect of the problem of quality selection is understandable. Nineteenth-century economists were preoccupied with the problem of value. Products (corn, tea, etc.) were thought to be well enough described by being named; their exchange ratios were the great mystery to be explored. Price theory thus became the core of economics. Its deficiencies led to attempts at further refinement. The problems dealt with in the writings of Sraffa, Mrs. Robinson, and Chamberlin have been mainly those arising out of the role of quality as a determinant of output and price. As mathematical tools of analysis have been elaborated, economic theorists have found additional reason to restrict their attention to the study of the quantitative elements in a market economy: costs, prices, outputs. The result has been that theorists today remain almost wholly preoccupied with the economics of price; nonprice elements have been studied not so much for themselves as for the manner in which they affect price-output behavior and policies.

The priority of price theory, of course, has not been the sole reason for the lack of an adequate theory of quality. Plausible grounds for deliberately excluding analysis of quality selection from economic theory are not lacking. Economists have at various times argued that such matters lie outside the scope of economics. Some assert that the problem of choosing the particular variant that best serves the purpose at hand is a technical problem; only when two or more ends compete for the same resources does an economic problem arise. Others claim that product preference is a matter of taste—which is a fit subject for investigation by students of aesthetics, psychology, and sociology, but which economists must accept as ultimate data. Still others stress the idea that consumers' choice of quality and producers' quality variation are in large part random, impulsive, or intuitive, and are not susceptible of theoretical generalization. It has also been argued that the starting point of economic theory must be market behavior, since one cannot obtain data on what people want but only on how they act.

Answers to these arguments will not be attempted at this stage of our inquiry. Some of them, I believe, will be found to be easily answerable once a theory of quality determination has been constructed. Others point to real and vexing difficulties that will need to be overcome or outweighed if a theory of quality is to make a net contribution to economics.

3. THE NEED FOR A THEORY OF QUALITY

But even if a strong case can be made for excluding the determinants of quality, a still stronger case exists for including them.

Economic theory is, after all, a set of tools for studying how organized society employs its resources in attempting to satisfy given wants. Present-day theory does this only incompletely. It starts in the middle of the analysis instead of at the beginning. Its explanation of the operations of a market in an exchange economy begins by accepting as a datum the existence of a certain commodity which is already desired by buyers conscious of some want which they believe the commodity capable of satisfying. How did the commodity come into existence? How did the buyers come to believe their needs would be met by buying this commodity? The theory does not tell us.

Yet if the theory of a competitive market economy is to become a truly *general* theory, must it not explain the whole situation in which an economy finds itself? If it is competent to explain the structure of outputs and prices but must accept the structure of goods and services as an unfathomable mystery, can it claim to be more than a *partial* theory?

It is the writer's contention that the study of competitive exchange, in order to be significant, must concern itself with the *quos* as well as the *quids* of an economy's *quid pro quo* transactions—that is, not only with the prices paid but with the things for which the payments have been made.

If an analysis is to start at the beginning, it must regard the exchange process not as the marketing of a given set of goods but as the outcome of desires to satisfy a given set of wants. And since the degree to which wants are satisfied varies with variation in quality, the economizing of resources necessarily involves consideration of quality as well as price. These two elements in economic choice are in fact inextricably bound together; the buyer must weigh them in conjunction. It seems unreasonable to insist that they be separated and assigned to different disciplines. Moreover, if the central problem of economics is that of "economizing"—that is, obtaining the greatest advantage from the smallest outlay—meaningful statements cannot be made about economic efficiency so long as quality considerations are excluded by hypothesis.

Herein lies a major disadvantage of excluding quality determination; it limits the usefulness of theory for welfare economics and for questions of public policy. If the community's welfare is directly affected by the number and kinds of commodities available, by the degree of product heterogeneity in each market, and by the occurrence of innovational or experimental product variation—in other words, by actions that broaden or narrow or change the character of the range of alternatives open to buyers—then any statement of the conditions of maximum welfare is seriously incomplete unless it specifies the conditions under which an economy has exhausted all possibilities for improving welfare through quality variation.

Price theory draws a curtain that shuts off from view this important area of economic activity. It is forced to draw this curtain provisionally, since analysis of price adjustments requires holding the qualities of products constant. Yet for this very reason price economics can provide only an incomplete description of the economic conduct of firms and households. The other half of the picture is obtainable only by permitting quality to vary. Unless it is realized that the assumption of a given set of products of invariant quality is merely a provisional one, this technique of price theory leads to a seriously distorted view of the economic world.

Chapter III

FOUNDATIONS FOR A THEORY OF QUALITY

1. THE WAY TO PROCEED: SOME COMMENTS ON METHODOLOGY

ON what general lines should a theory of choice of quality be constructed? One thing is clear: buyer's choice of quality is unanalyzable so long as market preferences are regarded as ultimate data for the economist. If each and every product which differs from others in any slight detail must be labeled as a separate commodity (the demand for which is "explained" solely by the statement that it reflects the "tastes" or "wants" or "needs" or "preferences" of buyers, whose motives and methods of quality selection lie outside the range of economic analysis), the theorist has no basis on which to find a common denominator for the heterogeneous products in those markets which are characterized by product differentiation.

For our purposes, then, the analytical methods employed in the theory of consumer behavior, as developed and presented by Hicks, Schultz, and others, are inappropriate. This body of theory does not provide tools by which the economist can examine factors other than changes in relative prices that influence consumers' decisions in matters of quality. The preference field technique is capable of showing only how a given pattern of demand (indifference map) for products already in existence determines changes in output when relative prices change. It cannot explain how demand curves and consumers' indifference maps arise and change, nor what determines their shapes, nor how a given demand, once established, becomes reflected in producers' subjective estimates of demand. Nor can it illuminate economic conduct which is aimed at gaining or providing further knowledge or experience in the light of which consumers' indifference maps can be reappraised and perhaps altered. When indifference analysis is employed, the existing set of commodities and buyers' current attitudes toward them as expressed in market behavior are regarded as data, and changes in products and attitudes are regarded as autonomous; they are considered to be disturbing factors which alter the conditions of equilibrium, not as part of the struggle toward equilibrium. Within this analytic framework, consumer choice of quality cannot be *explained;* it can only be *reported.* In this respect the scope of the present theory of consumer choice is comparable to the work of the geographer, who takes the earth's contours and formations as data and concerns himself with map making and map using, rather than that of the geologist, whose task is to explain the contours and formations in terms of forces and conditions affecting them.

If choice of quality is to be explained and not simply reported, the varieties of a differentiated product must be regarded as variable means to some end system, not as ends in themselves; other data must be sought as the determinants of the variables. Quality selection must be viewed as a form of adaptation, of action aimed at achieving the most advantageous means to given ends. In the study of consumer behavior the economist must penetrate the area that lies behind the curves and indifference maps of conventional theory, in order to relate choice of product quality to the consumer's more fundamental economic objectives. One prerequisite of a theory of consumers' selection of quality is that choice by consumers is to some extent deliberative, purposive, and rational. If the selection of a particular variety of product from among a collection of heterogeneous alternatives were something done wholly at random, or in response to impulses that must forever remain inscrutable, our task would be hopeless. Yet even if action is rational, a theory is not possible if the only thing known about consumer conduct is that it is consistent; for then we can do no more than decide that consistent patterns of behavior mean rational seeking of constant ends while seemingly inconsistent patterns mean either shifting ends or irrationality of conduct; we are still accepting these patterns as data, and must still restrict ourselves to analysis of the results of the conduct rather than the conduct itself. Unless ends *other than the possession of combinations of currently produced products* are specified, consumer choice of quality cannot be regarded as an economic act—that is, as a form of maximizing behavior. What is being suggested is that by the device of redefining the economic ends which underly demand the time-honored principle of economic analysis, the maximization principle, be applied to problems from which it has heretofore been excluded.

According to this view, there are two further prerequisites to a theory of choice of quality: a firm's product must be considered, not as the embodiment of some end, but as a variable means to some end; and the given ends of consumers must be defined in such a way that they do not necessarily coincide with the existing set of products—that is, correspondence between the ends sought and the means for attaining them must be capable of being imperfect and variable. Otherwise, product selection would involve no problem of conduct and would require no theory.

These prerequisites set the task before us. An acceptable theory must make clear, first, what the ends are, and, second, in what respects the relationship between ends and means is a variable one.

2. SOME USEFUL APPROACHES TO A THEORY OF QUALITY

Before we undertake the task of putting together a theory of quality determination, it will be helpful to examine some ideas put forward by economists which seem to suggest promising approaches to an analysis of the role of product quality in competitive markets.

One important idea is the distinction between buyers' preferences and buyers' patterns of purchases. A few writers have indicated—rather casually, it is true—that "tastes" and "wants" refer to something other than consumers' current attitudes toward the existing products as expressed in market behavior. Chamberlin, for example, speaks of the "problem of adapting the product more exactly to the buyers' (real or fancied) wants" and of wants being "more exactly satisfied with a differentiated product." [1] Wallace has referred to product differentiation as "a symptom of greater precision in the satisfaction of wants" and to "those sorts and degrees of variation of product which better meet the wants of consumers." [2] The language of these statements does not accord with the concepts and definitions of accepted demand theory; it implies that the product actually purchased by a consumer may provide less satisfaction (or perhaps less per dollar of outlay) than some ideal product which might conceivably have been selected.

Two questions now arise. To what is the want attached, if not to a specific commodity? And why should the discrepancy exist between the satisfaction actually obtained and the optimal satisfaction sought by the consumer?

A possible answer to the first question is provided by Frank H. Knight's statement that "What men want is not so much to get things that they want as it is to have interesting experiences." [3] The same idea is developed in Ruby T. Norris's *The Theory of Consumer's Demand.* According to Mrs. Norris wants should be thought of not as desires for goods, but rather for the events which the possession of them makes possible. "The economist misrepresents mankind when he supposes that paralleling the highly unreal *lines* of goods are just so many wants. . . . Goods are wanted because they are capable of performing services—favorable events which occur at a point in time." [4] Goods are in a relation of substitution when they are regarded as capable of performing some common service. Some goods may perform a given service with greater competence than others.[5] Here, then, is a framework of analysis that permits us to view a heterogeneous collection of products as variable means to some specified end.

A similar approach to the theory of consumption has been made

1 Chamberlin, *Monopolistic Competition,* pp. 10, 115 *n.*

2 Donald H. Wallace, "Monopolistic Competition and Public Policy," *American Economic Review: Papers and Proceedings,* XXVI (March, 1936), 81, 84.

3 Frank H. Knight, *Risk, Uncertainty, and Profit,* pp. 53–54. This quotation and others from Knight's *Risk, Uncertainty, and Profit* are reprinted with the permission of Houghton Mifflin Company.

4 Ruby T. Norris, *The Theory of Consumer's Demand,* pp. 136–37.

5 *Ibid.,* p. 78.

by James S. Duesenberry, in *Income, Saving, and the Theory of Consumer Behavior.* He begins with the premise that "people do not, for the most part, desire specific goods but desire goods which will serve certain purposes," that is, "to provide physical comfort or to implement the activities which make up the life of our culture." He continues: "Almost any activity can be carried out in a variety of ways and a variety of goods can be used to implement it. . . . The goods which can be substituted for one another for a single purpose . . . are qualitatively different ways of doing the same thing. Even more important they are not just different but some are better than others." [6]

Why, one may ask, does not free consumer choice lead to the elimination of all of the imperfect substitutes except the one best adapted to the purpose in hand?

One reason is that persons differ in their activities, partly because of differences in their aesthetic tastes and personalities, and partly because of differences in their situations—including such obvious elements as occupation, place of residence, and size of family. This reason is found in Chamberlin's statement that "Each 'product' has distinctive features and is adapted to the tastes and needs of those who buy it." [7] Mrs. Norris employs the concept of "commodity specialization"—the adjustment of the product to the purpose for which it is wanted—and mentions the fact, by way of illustration, that high-heeled dress shoes are imperfectly adapted to mountain climbing.[8] Duesenberry likewise points out that "When goods are looked at as the means of carrying out activities, their quality clearly varies with the degree to which they are specialized to suit specific purposes." [9]

A further reason, analyzed by both Mrs. Norris and Professor Duesenberry, applies to durable goods and lies in the fact that the individual occasions on which a consumer may want to use a certain type of good are often dissimilar. "Wants are not . . . single and one-dimensional. They are a complex of desires for different serv-

[6] James S. Duesenberry, *Income, Saving, and the Theory of Consumer Behavior,* p. 20.

[7] Chamberlin, *Monopolistic Competition,* p. 81.

[8] Norris, *The Theory of Consumer's Demand,* pp. 131–32.

[9] Duesenberry, *Income, Saving, and the Theory of Consumer Behavior,* p. 22.

ices, each unique in time if in no other dimensions. Ideally, a person might wish to choose a different commodity to render each service." An all-purpose product does not meet needs perfectly. Hence the desirability of acquiring "complementary substitutes." [10] Duesenberry employs the same reasoning: "A woman wants a large stock of dresses, not only for variety but because . . . if she owns only a few she is doomed to wear the same dress for different purposes." [11] Whenever the number of different situations in which a given class of product is to be used exceeds the number of varieties that can reasonably be offered on the market, adaptation must be imperfect, and the problem of quality determination becomes one of minimizing the unsuitability of the products.

A systematic study of this very problem has been attempted in the theory of spatial competition. Differences in consumers' scales of preference, due to differences in the environment in which they wish to use a given product, may be represented by differences in location of the population, while differences in the qualities of the varieties offered may be represented by differences in the location of the sellers. The number of sellers is assumed to be less than the number of populated locations; each variety of product therefore is "perfectly adapted" only to those who live in that precise spot. The inconvenience of obtaining a product not perfectly adapted to one's purposes may be represented by the incurring of transportation costs. In this way product variation can be shown as a movement along a scale. Hotelling and others have constructed one-dimensional models from which conclusions have been drawn as to the patterns of product differentiation to be expected under various assumed conditions. It is not necessary for us to examine these models at this point; it is sufficient to realize that here is a type of analysis which, if used with sufficient caution, is of great potential usefulness in studying quality determination.

Another approach sometimes used consists in assuming imperfection of knowledge but in treating the degree of imperfection as a variable. Consumers who are ignorant may act in such a way as to diminish their ignorance. Producers may also act deliberately in

10 Norris, *The Theory of Consumer's Demand*, pp. 130–51.

11 Duesenberry, *Income, Saving, and the Theory of Consumer Behavior*, p. 21.

ways to diminish the imperfection of their own knowledge and also to diminish (or augment) that of consumers. Few attempts have been made to introduce such hypotheses into formal analysis, yet they have occasionally been inserted. Chamberlin's analysis of selling costs is based partly on the idea that demand curves are altered when "the satisfaction of *existing* wants is sought with different information at the disposal of the buyer, as to the means whereby it may be done." [12] Reder has devoted one portion of his *Studies in the Theory of Welfare Economics* to an examination of the consequences of the explicit assumption that tastes are constant but that the knowledge requisite to the satisfaction of the tastes varies.[13] Mrs. Norris has placed considerable emphasis on the type of consumer's outlay which she calls "experimental purchases"—the buying of goods, typically in minimum amounts, for the purpose of finding the way, through trial and error, to new and improved patterns of spending.[14] Georgescu-Roegen has introduced a somewhat similar concept into the pure theory of consumers' behavior, with the hypothesis of a "penumbra of demand"—a range of indeterminateness of choice, which, for any given set of commodities, is reduced as time goes on as a result of the consumer's continued experimentation and consequent greater familiarity with them.[15]

This kind of approach opens up the whole broad question of the basis of choice and of methods of guidance in a world of imperfect knowledge. Long ago [16] J. M. Clark called attention to the fact that formal theory has ignored the processes of guiding and forming economic choices, though these processes must certainly be classed as productive since they are ways of giving things increased power to satisfy wants. The result, he asserted, has been "a type of theory that stops short of a full study of all the significant facts." Professor Clark mentioned particularly salesmanship as an example of "the shaping, informing, and guidance of human impulses which are responsible for their taking the form of definite desires and attach-

12 Chamberlin, *Monopolistic Competition*, p. 119.

13 Melvin W. Reder, *Studies in the Theory of Welfare Economics*, p. 69.

14 Norris, *The Theory of Consumer's Demand*, pp. 105–6.

15 N. Georgescu-Roegen, "The Pure Theory of Consumer's Behavior," *Quarterly Journal of Economics*, L (Aug., 1936), 569–71, 586–87.

16 In 1918. See John Maurice Clark, "Economics and Modern Psychology," *Journal of Political Economy*, reprinted in *Preface to Social Economics*, pp. 128–31.

ing themselves to definite concrete objects." Since he wrote those words considerable work has been done on the economics of sales promotion, yet analysis in this field has been almost wholly a study of predatory and monopolistic practices rather than an attempt to build a general theory of guidance. A full examination of the complex role of guidance in an exchange economy would have to include study of the kinds of guidance sought by producers who are faced with the problems of product variation in an uncertain world, and of the ways in which consumers attempt to guide themselves, as well as of the methods (so emphasized in economics textbooks) by which producers try to guide consumer decisions.

Joseph A. Schumpeter's theory of economic development also furnishes clues to a constructive approach to a theory of quality. Economic development, according to Schumpeter, consists of doing new things with an economy's resources—creating new products, adopting new methods of production, opening new markets, establishing new sources of supply. It therefore involves, among other things, the selection of heretofore untried combinations of qualities in consumers' and producers' goods. How are these qualities selected?

Schumpeter denies that the selection of new products is a routine process of adaptation to changes in buyers' wants, arguing that "the spontaneity of wants is in general small" and that "innovations in the economic system do not as a rule take place in such a way that first new wants arise spontaneously in consumers and then the productive apparatus swings round through their pressure." [17] In his model people's tastes and wants are given. Their horizons are limited. They are reluctant to change their ways. Yet the existing set of products is not assumed to be ideally adapted to their current needs; it is always possible to discover some new product that serves a purpose now being served by existing products and does it better, or more cheaply, or both. If a producer has sufficient courage to embark upon an untried project he will market the new product. Consumers do not at first want it. However, if it is actually superior they can be taught to want it and will eventually prefer it. Thus

[17] Joseph A. Schumpeter, *The Theory of Economic Development*, p. 65. See also his *Business Cycles*, Vol. I, p. 73.

consumer education is part of the process of economic development.[18]

It is worth noting that Schumpeter distinguishes between "wants" —that is, desires for particular products currently being produced —and "needs" which can be more or less well satisfied by these products, but could alternatively be satisfied, perhaps equally well or even better, by other products not yet made or wanted. This distinction makes it conceptually possible to classify products which are qualitatively different as alternative means to the satisfaction of a single "need." Innovation in the realm of consumers' goods can then be described as the creation of "a new good" which satisfies a need "better" or "more adequately." [19] According to Schumpeter's scheme, people's "wants" are likely to be to some degree out of step with their "needs." The preferable way to business success is to cater, not to "wants," but to "needs," for if an entrepreneur creates "a new good which more adequately satisfies existing and previously satisfied needs," [20] the mere existence of the superior good will tend to cause wants to change in its favor (not at once, it is true, but after a lag); while if he passes up such an opportunity so as to cash in on current "wants" his position is more precarious, since some rival may make the innovation and undermine his market. Thus the ultimate parameters of Schumpeter's system are not consumers' current indifference curves—these are variable through innovational activity—but the indifference curves that consumers would have in some millennium when all possible improvements had been made. Here lies the eventual, final solution of the grand economic problem of maximization of consumer satisfaction. Yet the solution can be achieved only by the method of successive approximations, and turns out to be endlessly slow and laborious, being at once a creative act that requires a touch of genius and a disruptive act that upsets the price-cost-output equilibrium and necessitates painful, time-consuming readjustments.

18 Schumpeter, *The Theory of Economic Development*, ch. ii; *Business Cycles*, Vol. I, ch. iii.

19 Schumpeter, *The Theory of Economic Development*, pp. 134, 135. Schumpeter also employs the term "purpose" to convey substantially the same idea, referring to "the case of replacing one production or consumption good by another, which serves the same or approximately the same purpose, but is cheaper."

20 *Ibid.*, p. 134.

Also noteworthy is the idea embedded in Schumpeter's theory that knowledge is incomplete, and horizons are limited. This idea is necessary, in the first place, to account for the imperfect relationship between consumers' "wants" and "needs." But Schumpeter emphasizes it especially as an inescapable condition in the business world. "As military action must be taken in a given strategic position even if all the data potentially procurable are not available, so also in economic life action must be taken without working out all the details of what is to be done. Here the success of everything depends upon intuition, the capacity of seeing things in a way which afterwards proves to be true, even though it cannot be established at the moment." [21] Schumpeter stresses the uncanny knack of being correct rather than the elimination of incorrect courses of action through diligent experimentation and learning. This, however, is less important than the fact that his theory requires, as one of its essential conditions, a state of imperfect knowledge.

In short, Schumpeter's theory provides an approach fairly similar to that pieced together from the ideas of Duesenberry, Norris, and Clark. Implicit in it are the propositions that products are not ends of economic activity but variable means to other ends, describable as the "purposes" which the products serve, or the "needs" they satisfy, and that lack of omniscience on the part of producers and consumers explains why the means are variable even when the ends are fixed. What is chiefly lacking is a detailed analysis of how consumers and producers act in order to bring the means in closer conformity to the ends.

These ideas suggest the kind of theoretical framework that is needed for a study of the determinants of product quality. Concepts similar to them will be elaborated in the chapters that follow.

First, consumers' objectives will be defined as something other than the purchase of various amounts of currently produced commodities. The economy's products will be regarded as variable means to some predetermined set of ends. These ends may be described as the events, activities, and experiences in connection with which consumers' goods are used. Quality variation and consumer choice of quality will then be looked upon as forms of adaptation.

[21] *Ibid.*, p. 85.

Second, it will not be assumed that such adaptation is perfect and effortless—that the products bought by consumers correspond perfectly with the ends they are meant to serve. "Wants" and "tastes" cannot be considered as synonyms for "market preferences." Instead, the fact of imperfect correspondence between means and ends, and the consequent variable relationship between the two, will form the basis of the theory. The imperfection of correspondence will be regarded as due in part to the diversity and the constantly changing character of people's tastes, needs, circumstances, and environments; in part to their imperfect knowledge, and hence to the opportunity for improvement through continual investigation and experimentation; and in part to the fact that the ends of people as producers conflict with their ends as consumers, setting up cross currents of activity which call for explanation.

Our central concern will be with the consumer, but the proposed method of analysis will embrace both consumers' choices between available qualities (the limits being set by what producers choose to offer) and producers' choices between possible qualities (the limits being set by the existing technology and entrepreneurial ingenuity).

Part Two

THE THEORY OF CHOICE OF QUALITY

Chapter IV

BASIC AND DERIVED WANTS

SOMEWHERE behind the market demands for goods and services in an economy lie the ultimate ends of its individual members—ends that may be sought at least in part through economic action. These ends are not the same as the products purchased. The buyer of electric current wants something other than the mere fact of current passing through a copper wire; he wants, perhaps, illumination by which to read a magazine, or heat that will turn bread into toast. And it is likely that he does not want the slice of toast or the ideas in the magazine for themselves but as means to something else. Products are linked to human aspirations, desires, and impulses. In this chapter an attempt will be made to show what lies back of demand—to describe the nature of those "basic" wants that underlie wants for specific products, and that therefore are the mainsprings of economic conduct.

There are risks involved in this approach. It requires one to probe into matters that lie beyond the measuring rod of money. It invades the fields of psychology and philosophy. It calls for generalizations based on "subjective" rather than "objective" observation. These are risks worth incurring, however, if the approach provides a firmer basis for comprehending and appraising the all-over operation of an economic system. Succeeding chapters, it is hoped, will convince the reader that it does.

1. BASIC WANTS

The thesis of this section may be stated quite simply. What people really desire are not products but satisfying experiences. Experiences are attained through activities. In order that activities may be carried out, physical objects or the services of human beings are usually needed. Here lies the connecting link between man's inner world and the outer world of economic activity. People want products because they want the experience-bringing services which they hope the products will render. Two levels of wants are thus distinguishable. The more fundamental kind of want—the desire for an experience—will be termed a *basic want;* its derivative—the desire for a product which actually or supposedly provides the means to that experience—a *derived want.*

This thesis in no way implies dependence on the now discredited hedonistic psychology. Some people, especially among those living in Western nations, may want pleasure or happiness, or may want it some of the time; but a general explanation of conduct cannot rest on so limited and incomplete a foundation. Experiences of other sorts are desired. People crave power; in fact the late Frank D. Graham contended that the single generic term power "all but completely covers human aspirations." [1] Various people also want at various times glory, distinction, adulation, contentment, psychological security, oblivion—or even the privation and suffering that is believed by some to be a prerequisite of salvation. Some wants are based on altruism; Marshall mentioned "the desire to do one's duty" and the fact that one may "rejoice in another's joy." [2] Sociologists and psychologists sometimes speak of tensions and release rather than wants and satisfaction.[3] This terminology is perhaps better suited than the economist's customary language to describe the grim satisfaction of an avenger. It is certainly appropriate when dealing with some of the drives and urges which have a physiological basis. C. Reinold Noyes, in his monumental study of the possible physiological roots of economics, describes wants in terms of deviations from "homeostasis," which may be roughly described as a condition of bodily equilibrium, with its accompanying freedom from physical or mental discomfort. Thus man's wants are "states that he wishes to get rid of." They include hungers, appetites, temperature ex-

[1] Frank D. Graham, *Social Goals and Economic Institutions*, pp. 12–13.

[2] Alfred Marshall, *Principles of Economics*, pp. 16, 17 *n.*

[3] See, e.g., Lawrence K. Frank's "The Management of Tensions," *American Journal of Sociology*, 1928, pp. 705–36, reprinted in his *Society as the Patient*, pp. 115–42.

tremes, disagreeable tastes, ugliness, disliked effort, inconvenience, tedium, etc. In such instances what we customarily call "satisfaction" seems best described as relief from unpleasantness.[4]

Another aspect of wants, now well recognized by economists as well as sociologists, is the fact that they are largely determined by one's culture. Most wants we feel not as isolated individuals but as members of a social group guided largely by customs, conventions, and taboos. Most of the activities we engage in are thus determined. Often we "want" not what we really desire but what we think we ought to want.[5] Desire for the approval or envy of others prompts many actions that otherwise seem to run counter to rational calculation. Even in the satisfaction of physical wants people are guided in considerable measure by their ideas of what is "being done"; as Malinowski has expressed it, "Appetite or even hunger is determined by the social milieu." [6] Not wholly so, of course. Just as physical urges are modified by cultural pressures, so these motivating forces are modified by one's individual concepts and standards.

Complex and diverse as the bases of men's wants thus appear to be, the general statement that "what people desire are satisfying experiences" nevertheless remains true. What is considered satisfying is a matter for individual decision; it varies according to one's tastes, standards, beliefs, and objectives—and these vary greatly, depending on individual personality and cultural environment. Here is a foundation for a theory of choice broad enough to embrace Asiatic as well as Western cultures, nonconformists as well as slaves to convention, Epicureans, Stoics, Cynics, roisterers, religious fanatics, dullards, and intellectual giants alike.

[4] C. Reinold Noyes, *Economic Man in Relation to His Natural Environment*, I, 29–42, 408–9.

[5] See Frank H. Knight, *The Ethics of Competition*, p. 22.

[6] Bronislaw Malinowski, "The Group and the Individual in Functional Analysis," *American Journal of Sociology*, XLIV (May, 1939), 943. See also Knight, *The Ethics of Competition*, p. 26.

2. VALUE SYSTEMS AND CONSTELLATIONS OF WANTS

How do people evaluate the possible experiences open to them, in order to decide whether or not to seek them? According to Talcott Parsons, "A rational process of allocation involves not merely given ends, but a coherent system of ends, a scale of relative valuations. . . . In other words each individual's actions must be thought of as in the last analysis subordinate to a set of ultimate ends or principles of action." [7] That is to say, each person must have his own set of values or standards by which to measure the relative desirability—for him—of different experiences. Such a value system is needed because unlike experiences have nothing in common—nothing scientifically demonstrable, that is—by which to measure and compare them. Unless, therefore, individuals evolve their own standards of comparison experiences must remain incommensurable, and choice is impossible.

A value system is needed not only to determine whether or not to undertake some contemplated activity but also to determine, among the various possible ways of doing it, which is to be preferred. This is so because every change in the detailed manner in which an activity is carried out alters somewhat the experience obtainable from it. A "want" for an experience is in truth not a single want but a complex of related, supplementary wants, usually consisting of a major want plus numerous minor wants. Following Noyes,[8] I shall employ the term "constellation of wants" to denote such a clustering.

An example may help to clarify this point. A man uses a razor or electric shaver because he wants to enjoy regularly the experience of being beardless and well groomed. The activity required in order that this end be accomplished is shaving. But shaving inevitably touches other basic wants besides the want for beardlessness. It may cause pain or irritation, and so conflict with the desire

[7] Talcott Parsons, "Some Reflections on 'The Nature and Significance of Economics,'" *Quarterly Journal of Economics*, XLVIII (May, 1934), 517.

[8] Noyes, *Economic Man*, I, 183–85, 404–10. See also John Bates Clark, *The Distribution of Wealth*, pp. 235–44, and Frank H. Knight, "Realism and Relevance in the Theory of Demand," *Journal of Political Economy*, LII (Dec., 1944), p. 298.

to avoid discomfort. It is occasionally troublesome, as when repairs or new blades are needed. It consumes time, so runs counter to the desire to minimize the time spent on dull, routine matters. Beardlessness, too, is a matter of degree; if the razor fails to shave closely or uniformly the satisfaction derived from a shave may fall considerably short of what had been anticipated. Here are a number of incommensurable variables, each associated with a different want. The satisfactoriness of a razor depends on the way in which the *combination* of wants is satisfied.

How does one weigh a loss in one respect against a gain in another? The only possible method is by a personal system of values. Some men may attach great weight to timesaving; others may be indifferent to the time element but fret about the slightest facial discomfort. The precise weighting of the various wants in a constellation must be dictated by the underlying tastes and preferences of the person concerned. They are the ultimate data which determine consumers' choices.

Every activity similarly involves a constellation of wants and is subject in the same way to innumerable modifications in the degree of satisfaction accorded to each of the various wants in the cluster.

The term *constellation of basic wants* may be employed to describe the totality of wants relating to some experience or activity. The term may also be employed in a more limited sense. Typically activities require the services of several products used jointly. (In tennis, for instance, one needs a court, net, racket, balls, and proper shoes and clothes.) A desire for an experience naturally implies a desire for all the needed services contributed by all the component products in the group. But if the word "want" is employed in its literal sense, meaning not simply need or desire but also lack, a person who possesses all but one of a set of complementary products cannot be said to want all of them, but only the one that is missing. In such a case he goes into the market with a limited set of basic wants. Ordinarily the phrase "constellation of wants" will be used in this restricted sense, to refer to those particular wanted services and satisfactions which a single product, having a well-defined function when used jointly with others, is equipped to provide. Usually the kinds of goods needed in connection with wanted activities are definite in number and character, these being determined either by convention or by the existing technology. Job activities require conventional clothing, entertainment of dinner guests necessitates acquisition of certain accepted sorts of foods and beverages, etc. It is therefore possible to consider the wants appropriate to each as separate constellations. For instance, the wants for uncooked food, for cooking containers, for stoves, and for sources of heat may be considered as derived from separate and distinct clusters of wants, each associated with a complex of services which products of a certain type are capable of rendering.

One complication arises in connection with the concept of a constellation of wants. People's wants for a given kind of activity or experience may vary widely in their generality.

A person may want the experience of eating food of some sort, without having any preference as to its nature; or he may want the more specific satisfactions associated with meat; or his want may be still more specific and include a yen for the flavor of roast beef. In the first instance the pangs of hunger are of overriding importance; in the second, some subsidiary wants are strong enough to rule out certain types of food as incapable of bringing satisfaction; in the third, an added member want in the constellation is a desire for repetition of a particular taste experience.

It seems correct to say that the more vague and general the desire is, the fewer are the component wants in the constellation, the weaker are the subsidiary wants, and therefore the broader the range of different products which can be considered as alternative means to the desired end. It is well to recognize this complication, and to be aware that a product of given specifications may be a means to the satisfaction of various constellations of wants, though perhaps able to serve some more successfully than others.

Because of the plurality of wants in a constellation, incidentally, choice of quality is an economic act even within the limited meaning of the term "economic" as laid down by Lionel Robbins. For the choice of any variety of a product *does* involve the apportionment of resources among several competing ends—the various component wants of a constellation—each of which may be accorded varying degrees of satisfaction.

Failure to realize that products usually serve not only some main purpose but also a number of ancillary purposes, and that each individual's attitude toward the relative importance of each of these purposes is a value judgment which may be open to criticism on aesthetic or ethical or physiological grounds but not on economic grounds, has led writers to make statements that confuse rather than clarify. Even Knight seems to have fallen into this error in his earlier writing. In *Risk, Uncertainty, and Profit* he referred to "elements in the physical form and appearance of a commodity which make no difference in its efficiency for the purpose intended (an agreeable color, decorative ornament often actually interfering with its uses, fancy containers, etc.)." [9] According to this statement the efficiency of a product is to be measured solely with regard to the degree to which it satisfies the main want. Since the chief function of a toothpaste is to clean teeth, and of meat is to provide nourishment, it would follow that a disagreeable flavor in no way impairs their efficiency. Yet flavor cannot always be thus disregarded. In the case of candy or an after-dinner cordial, whose chief function is to provide an agreeable flavor, the product's "efficiency" must depend directly on its flavor. This sort of reasoning leads to difficulties. Desserts obviously serve at least a dual purpose; providing both nourishment (or needed elements in a balanced diet) and a delectable flavor. To select one of these two functions as "the purpose intended" and regard the other as an element "which makes no difference to its efficiency" would be wholly arbitrary. And what shall we say about a decorative vase, one that graces a living room even when it is empty of flowers? Can the very shape that gives it beauty be said to "interfere with its uses"? The idea that efficiency is concerned only with practical uses, and not with aesthetic ones, is in itself a value judgment.

Another clearer instance of the rather common failure to perceive the plurality of uses of a product is to be found in a passage in John S. Gambs' *Beyond Supply and Demand* describing Veblen's attempted distinction between "the essential function of a commodity and the perverted function (e.g., women's shoes: essential function to protect the feet and facilitate walking; actual and per-

[9] Frank H. Knight, *Risk, Uncertainty, and Profit,* p. 157 *n.*

verted function, to deform the feet and inhibit walking)." [10] Actually women's shoes serve several purposes, among which are to make their wearers look attractive and stylish. What is described as a perverted function is actually an unavoidable nonmonetary cost —a sacrifice of some amount of one kind of satisfaction in order to obtain more of another kind. And what is evidently objected to is a person's having a set of values which rates beauty and stylishness above maximum comfort and health.

3. DERIVED WANTS

The connection between *basic wants* (wants for experiences) and *derived wants* (wants for products) is easily established.

Usually it is impossible to satisfy a basic want without the aid of some appurtenances in the form of physical objects or personal services. In their efforts to satisfy basic wants, therefore, people develop wants that become attached to particular products. It seems appropriate to term these "derived wants" since products are not wanted for their own sake, but only because they are associated in the buyers' minds with certain satisfactions or sets of satisfactions; and it is of course the satisfaction, not the product, that is really wanted. Hence basic wants may be said to indicate the ends of economic activity (though perhaps not ultimate ends), while derived wants indicate the provisional choice of means to those ends.

A derived want presupposes an acquaintanceship with the product. As Marshall has observed, "the demand for a thing depends in a great measure on people's familiarity with it." [11] Basic wants, on the other hand, are for the most part independent of products or markets, though in some cases new products bring with them new experiences and thus create new basic wants, and it is also frequently true that window displays and advertisements often remind people of basic wants that have lain submerged.

In order for a want-for-satisfaction to be converted into a want-for-a-product, several things must occur. First, the person must be-

[10] John S. Gambs, *Beyond Supply and Demand*, p. 20.

[11] Marshall, *Principles of Economics*, p. 486.

come aware of the existence of some product capable, more or less, of satisfying the basic want. Then it must somehow dawn upon the person's consciousness that the product *will* satisfy his want—which means awareness not only of the product's existence but also of the uses to which it can be put. And then the person must mentally weigh the advantages of the product against those of whatever alternative products occur to him as being capable also of satisfying his basic want, and thus establish some scale of preference. Only then does the individual's demand curve take form.

How useful is the distinction between the two kinds of wants? It has been asserted that "it is a matter of no concern whether we want the things or the conscious states we expect to derive from them."[12] This would be true only if the things purchased always produced the desired states. But in the actual world products are typically imperfectly suited in some respects to the wants they are intended to serve; in many instances this is bound by the very nature of things to be the case. And whenever a variable degree of imperfection exists, the concepts outlined in this chapter are crucial to the study of that portion of economic conduct which is aimed at maximizing the degree of suitability.

[12] Knight, *Risk, Uncertainty, and Profit*, p. 60 *n.*

Chapter V

SOME NEEDED ASSUMPTIONS

IN constructing a theoretical model of an economic system it is necessary to decide what conditions are assumed to be present or absent; only when the assumptions are made explicit can the precise operation of the model be determined. This chapter will be concerned with the assumptions to be made in analyzing consumers' and producers' behavior in a market economy when faced with the problem of choice of quality. It is desirable that we assume conditions which reflect the essential features of the real world, yet we must also bear in mind that the quest for realism may lead us into complications that are unmanageable. We must somehow strike a balance between simplicity and conformity to reality.

1. DISSIMILARITIES IN CONSTELLATIONS OF WANTS

Shall we assume homogeneity or heterogeneity of population with respect to basic wants? In other words, which is the more essential fact—the similarities in people's needs, tastes, circumstances, and activities, or the differences?

In the recent literature both assumptions are to be found. Knight's classic analysis of perfect competition includes the assumption that "The members of the society are supposed to be normal human beings . . . differing among themselves in the ways and to the degrees familiar in a modern Western nation." [1] On the other hand, Samuelson finds it "characteristic of much of modern thought" to assume that "the welfare function is completely (or

[1] Frank H. Knight, *Risk, Uncertainty, and Profit*, p. 76.

very nearly) symmetrical with respect to the consumption of all individuals," which implies that individuals are "very much alike, and if given equality of treatment would develop the same want patterns." [2] It must be added that he is critical of this thinking on the grounds that "it is not consistent with the patent fact of considerable differences in individuals' overt preference patterns."

What are the facts?

In the first place, dissimilarity of tastes and values is an inherent, inevitable feature of human society. People are bound to differ in age, in maturity of outlook, in their stages of personality development, and in physical characteristics. They differ too in other unfathomable ways. Disagreements about religious beliefs, methods of arranging a living room, personal qualities of mutual acquaintances, the best way to spend a vacation, the right amount of sugar and cream in coffee, serve to illustrate the infinite diversity of human likes and the impossibility of finding tastes that are identical in every last detail. Even people who are alike in other ways may differ in their ability to discriminate finely.

Second, people's environments and circumstances differ significantly in numerous ways. Some have larger incomes than others. Some live in urban, others in rural communities. Locations of homes differ with respect to distance from shops, the railroad station, etc., resulting in differences in transportation needs. A person's occupation affects his wants for clothing. Even if all people were identical in their value systems, their buying patterns would differ.

A third reason for the diversity of wants lies in the fact that each person pursues his activities under variable circumstances and in a changing environment. Changes in the weather alter wants. The occasions on which products are used differ. Variable wants may be served either by a single product that has the virtue of adaptability, such as an all-purpose tool, or by a number of specialized products. Even if a population were homogeneous in its tastes and circumstances, products would have specialized uses, and wants for them would not all be identical.

For still another reason a person's wants are likely to be different at different times: variety is wanted for its own sake. Novelty is

[2] Paul A. Samuelson, *Foundations of Economic Analysis*, pp. 224, 225.

often sought as an end. One's ranking of flavors of ice cream is likely to be a function of the flavors most recently experienced. As Frank Knight has told us, "Man's chief interest in life is after all to find life interesting, which is a very different thing from merely consuming a maximum amount of wealth. Change, novelty, and surprise must be given a large consideration as values *per se*." [3] Even if people were homogeneous in their tastes, their efforts to gain diversity of experience by deliberately selecting one variety of product after another, either in rotation or by progressing to ever new varieties, would produce precisely the same effect in markets as if they had different tastes.

True, there is a contrary tendency toward similarity of activities and wants. People wish to be like those they admire: affluent neighbors, stage heroes, parents—in general, the "best people." [4] They hate to be thought queer or abnormal, or are reluctant to try the unknown, and so prefer what is common, familiar, and accepted. Even people whose tastes deviate from the crowd's in minor respects may find it less troublesome to conform to standard usage. In the case of "standard" items such as bicarbonate of soda and stove coal, people are usually oblivious of quality so long as it is not appreciably "below standard." Yet these factors do not wholly offset the tendency toward diversity; they simply make the differences less pronounced, and changes less rapid, than they would be otherwise. The essential fact remains that people are dissimilar in their basic wants, and that the differences are pervasive and significant. In fact, people's patterns of desires and standards are so many, and so varied, and subject to such infinite gradations, that it seems nearer the truth to conceive of each individual as being slightly different in at least some respects from all other individuals.

In studying certain kinds of problems it may be quite legitimate to assume homogeneity of basic wants for purposes of simplification. In investigating the behavior of the savings ratio, for example, the theorist's concern may be with aggregates or averages at each of various income levels, and not with the dispersion around the

[3] Knight, *Risk, Uncertainty, and Profit*, p. 369.

[4] Cf. Otto Klineberg, *Social Psychology*, pp. 319–44. But of course the affluent neighbors then seek new ways in which to be different.

averages; he may deliberately disregard individual disagreements in matters of taste in order to reveal more clearly the factors responsible for differences in the behavior of typical or average households at different levels of income.[5] But in a general analysis of the exchange process and of market structures and institutions, neglect of individual differences can seriously falsify the conclusions. Imagine how far astray a political scientist would be led if he attempted to construct a model of a political system on the assumption that all citizens were alike in their interests! There is a similar danger that economic analysis which ignores diversity of wants will obscure rather than reveal the ways in which buyers' and sellers' behavior and market institutions and structures in an exchange society are related to people's needs and wants.

In short, diversity of basic wants is so significant a feature of our modern economic world that to assume the contrary is to assume away one of the main reasons for the behavior patterns and market structures economists wish to explain. Accordingly, it will be assumed that sellers in each market are faced with a very large number of different constellations of basic wants, and that the differences are substantial and important. It is not necessary to assume that every buyer in a market has wants that are to some slight degree different from those of any other buyer; however, this may be a convenient form in which to put our general hypothesis.

2. IMPERFECT KNOWLEDGE OF UTILITY

Shall we assume that all participants in the exchange process have perfect knowledge? Or is it better to assume that consumers are uncertain about the capacity of products to satisfy their wants, and that producers are uncertain as to what wants consumers are seeking to satisfy, and therefore as to the desirability of their products in the minds of potential purchasers?

In the latter case we are dealing with imperfect knowledge of utility—that is, of a product's ability to perform a desired service.

[5] See James S. Duesenberry, *Income, Saving, and the Theory of Human Behavior*, especially pp. 20–22.

To know the utility of any product one must know both the character of the service desired and the specifications of the product. Consumers are more apt to know the former than the latter; with producers it is likely to be the other way round. If one's knowledge of either is in any way incomplete, knowledge of utility is necessarily imperfect.

Orthodox value theory generally takes knowledge of utility for granted. Marshall chose this course when he selected "the price which a person is willing to pay" as the measure of desire, and decided to "make it serve, with all its faults, *both* for the desires which prompt activities and for the satisfactions that result from them." [6] In recent years, it is true, a considerable body of analysis has grown up which deals specifically with conditions of uncertainty and imperfect expectations. Knight's theory of profit, Keynes's marginal efficiency of capital, Hicks's elasticity of expectations, Hart's treatment of anticipations and business planning, Shackle's expectational dynamics, and Tintner's analysis of consumer's choice under uncertainty are examples of theoretical work that assumes imperfection of knowledge. But in all of these cases, the uncertainty involved is uncertainty about quantitative matters: prices, quantities demanded, costs, interest rates, income. None of these writers has introduced the assumption of uncertainty as to the want-satisfying power of goods, or entrepreneurial uncertainty as to whether the qualities of his product are those best suited to his customers' needs.[7]

Yet the common facts of everyday life make it clear that the postulate of perfect knowledge of utility presents an outrageously false picture of human behavior. Utility springs into being only with the actual realization of an experience. No one can know in advance precisely what an experience will be like. Though new experiences frequently approximate previous experiences, they are as a rule different to some extent; and even when two satisfactions

[6] Alfred Marshall, *Principles of Economics*, pp. 92, 93 *n*.

[7] Chamberlin and other writers on monopolistic competition have drawn on this assumption to explain the phenomena of advertising and the selling of virtually identical goods under different brand labels, but have not used it to develop any formal theory of market behavior. Melvin W. Reder has taken a tentative step in this direction in his *Studies in the Theory of Welfare Economics*, p. 69, where he examines the consequences of the assumption that tastes are constant while the knowledge requisite to the satisfaction of tastes varies.

turn out to be indistinguishable, one cannot know in advance that they will be. Hence demands are based on *estimates* of utility.

There are countless reasons why estimates are likely to be faulty. Judgments before purchase are usually made on the basis of inadequate sampling or hearsay evidence or simply willingness to try. The colors of a necktie or dress selected under the lighting conditions in the shop where it is bought may look different afterwards. The tone quality of a radio delivered to one's home may differ, either actually or in imagination, from that of the demonstration model in the showroom. A wallpaper pattern may not give quite the effect anticipated from seeing a small sample. True, in the case of frequently repurchased goods—some foods, drugs, toilet articles, etc.—realized utility may often be identical with anticipated utility. But such goods form only a minority of consumers' outlays. In the general case, the exchange process can be accurately described only in terms of anticipations, realizations, pleasant surprises, and disappointments.

The seller is in the same situation. He cannot know with certainty the kind of experiences his customers prefer, the kind of anticipations they are likely to form, or the extent to which their anticipated satisfactions will be realized without disappointment. Therefore he too must depend upon estimates.

The importance of introducing the assumption of imperfect knowledge of utility into the theory of choice becomes clear when we reflect upon the fact that uncertainty as to the want-satisfying powers of goods introduces a qualitatively new dimension into economic behavior.

In economic theory the assumption is generally implied that the selection of any course of action is effortless, costless, involves no problems, and therefore can be ignored. The fact is, however, that doing something is the second part of a two-part process; pondering over what to do must precede it. This is true of small as well as of large decisions, of details in fairly routine matters as well as of strikingly novel courses of action. People must work out in imagination the consequences of alternative decisions. If needed information is lacking, they must go about acquiring it. Moreover, much choosing must necessarily be exploratory and experimental. The methods employed in scientific research—painstaking study, an attitude of skepticism, gathering of further data, retesting, building up an accumulation of experience—are, in a humbler way, also characteristic of everyday economic conduct. Hence an important characteristic of economic behavior is its continual modification in the light of new knowledge. This is simply an exemplification of the learning process. Mistakes inevitably lead to regrets and readjustments. Conjectures that prove to be erroneous are eventually discarded or revised.

In a stationary society composed of individuals whose ages, homes, occupations, and incomes never varied, experimentation would gradually eliminate the need for further experimentation. But in the world of reality, with its changing environment and progressive character, and hence continual formation of new wants, uncertainty about utility and trial-and-error methods tend to persist. All this suggests the desirability of constructing a theoretical model in which the assumption of imperfect knowledge of utility is included.

What we shall assume is that buyers have imperfect (that is, either incomplete or partly erroneous, or a mixture of both) knowledge of what goods are available in markets, what uses they serve, and how well they serve them, while sellers have imperfect knowledge of buyers' basic wants, of the manner in which variations in patterns of wants are distributed throughout the population, and of the particular qualities in a product that would best satisfy any given constellation of wants. We shall also assume that the degree of imperfection of knowledge (and therefore the state of expectations) is variable, and that means to alter the state of knowledge are available.[8]

[8] The necessity for assuming that the state of knowledge is variable is obvious: otherwise experimentation and other forms of activity aimed at bettering (or worsening) knowledge would be, *ex hypothesi*, doomed to failure. On this point see Joseph A. Schumpeter, *Business Cycles*, I, p. 55, and Paul M. Sweezy, "Expectations and the Scope of Economics," *Review of Economic Studies*, V (1937–38), 234–35. The limited usefulness of the cobweb theorem is due to its dependence on the assumption that expectations remain constant—in other words, that no one learns from experience.

3. FURTHER ASSUMPTIONS

In addition to the two main assumptions described above—dissimilarities in wants and imperfect knowledge of utility—seven more will be found useful in building our rudimentary model of consumers' and producers' behavior.

3. *All action is rational.* That is, every person and firm has a system of values and acts consistently in accord with it, making the most intelligent decisions possible in view of the handicaps of uncertainty and limited information.

Irrationality cannot be dismissed as an unimportant element in human behavior; no one, in truth, behaves altogether rationally. It is a bold move to abstract from all thoughtless, gullible, and inconsistent behavior. Yet the hypothesis of rationality is essential to economic analysis. When choice is made without method, conduct is unpredictable and analysis is impossible. Rationality really implies no more than the employment of some sort of method; it is perhaps best defined as the consistent use of a set of principles on the basis of which choices are made. In this sense, rationality does not exclude all forms of impulsive or emotionally motivated action; it is rational, for example, to make an impulsive choice if it is done deliberately, either for the sake of the mere joy of acting on impulse or for the adventure of experiencing consequences not carefully weighed in advance. Nor does it exclude decisions made when knowledge is incomplete; a finesse in bridge may be the rational play, even though its consequences turn out to be disastrous. What it does exclude is action based on erroneous reasoning or on incomplete calculation that fails to take into account all the readily available data or on the untempered dictates of one's emotions or on the unthinking acceptance of advice from others.

If we assume rationality we of course exclude an important element of human behavior. Yet there is more justification for doing so than for assuming perfect knowledge, for irrationality is not quite so pervasive or ineradicable as uncertainty. People who have highly developed intellects and have learned to subdue and control their emotions probably act most of the time in a way that roughly approximates completely rational behavior, whereas no one, however superior he may be, tends to acquire knowledge without expenditure of time and effort, or to be free from uncertainty as to the consequences of his major decisions. The rationality assumption is also a useful device to separate conduct which can be explained only by the existence of irrationality from conduct which arises from other causes, such as imperfection of knowledge.

4. *Consumers' value systems and circumstances, and hence their basic wants, remain constant.*

This assumption too is contrary to fact. Not only do people's wants for products continually shift and grow, but their wants for experiences undergo continual modification, partly due to changes in their circumstances, but also due to changes in their inner personalities and standards. It is in man's nature to develop in body, mind, and spirit.[9]

The fact that this is so, however, does not of itself invalidate the procedure of accepting ends as data. So long as tastes and values are unaffected by economic forces they may be treated as parameters, even though they do not actually remain invariant. Conclusions drawn from the assumption that they do are valid statements as to what *would* have happened if ends had remained unchanged.

Yet "*ceteris paribus* is a slippery tool," Arthur F. Burns reminds us,[10] "and may lead to serious error . . . if the impounded data are correlated in experience with factors that the theorist allows to vary, or if the very process of adjustment induces changes in the impounded data." The real danger to the validity of an analysis based on the assumption of constant values lies in the fact that values are in truth significantly affected by economic forces. Culture and personality are in part a product of economic institutions. Consumers' goals are altered by advertising and by the purchases of friends and neighbors. New products pave the way for new experiences which give individuals a new outlook on life. It simply

[9] Cf. Marshall, *Principles of Economics*, pp. 89–90; Frank H. Knight, *Ethics of Competition*, pp. 22, 101.

[10] Arthur F. Burns, "Economic Research and the Keynesian Thinking of Our Times," *Twenty-Sixth Annual Report of the National Bureau of Economic Research*, p. 8.

is not true that economic goals are independent of the means evolved for reaching them.

Nevertheless it is necessary for us to assume so. Only by this device of keeping ends fixed can we study an adjustment process in isolation and determine the final outcome.

Nor will the resulting errors be so great as to destroy the usefulness of the analysis. Basic wants are in fact relatively stable. Habit, inertia, and the desire for the familiar undoubtedly outweigh the yearning for the new and different. If consumers' tastes and values were largely capricious and unpredictable, modern large-scale production methods involving heavy fixed capital investment would be impractical and in fact no organized market economy would be feasible. So long as consumers' goals do in fact have enough stability to give consumers' behavior a somewhat predictable pattern our assumption is a workable one.

5. *Producers seek either to maximize profits, or to maximize their rate of growth (or, alternatively, their sales volume or share of the market), or some combination of the two.*

This assumption permits us to take into account the empire-building aspirations as well as the monetary appetites of business leaders. It is customary to assume that business firms seek to maximize monetary gains. Yet it is generally conceded that businessmen are driven by other motives which are perhaps equally or even more important.[11] Perhaps the truth is that business "values," like consumers' values, are plural, and that the goal of each business firm, depending on the character of its leadership, is a differently weighted combination of gains in sales, profits, power, and prestige. For purposes of price theory, an assumption that included sales volume as well as profits as ends of business conduct would greatly complicate the analysis (though it might be worth trying), since it would require the setting up of an indifference function showing how much of an expansion in sales volume would be needed to compensate for any given decline in profits. For our immediate purposes, however, no such difficulties are involved. And there is

[11] See Joseph A. Schumpeter, *The Theory of Economic Development*, p. 93; Robert A. Gordon, *Business Leadership in the Large Corporation*, pp. 305–12; John Maurice Clark, "Toward a Concept of Workable Competition," *American Economic Review*, XXX (June, 1940), 247.

the advantage that any theory of producers' behavior consistent with the assumption that either or both profits and size are ends of business conduct is a more general theory than one which rests on profit-maximization alone.

6. *Producers do not attempt to exploit buyers.* Hoodwinking, chicanery, and fraud are ruled out.

This, perhaps, could be left unstated, being quite possibly a corollary of the rationality assumption, but it seems best to make it absolutely clear that deceptive tactics will play no part in the model.

Admittedly, this is a step away from realism. In real-world markets, where uncertainty about utility is omnipresent, producers' efforts to exploit consumer ignorance through deceptive forms of product variation and deceptive selling methods are commonplace phenomena. Yet they are by no means universal. If, therefore, we wish to obtain a general picture of the competitive exchange process we must, at least at first, abstract from this kind of behavior.

7. *All products are consumer goods.*

This assumption is made merely for convenience, since it simplifies the description of buyers' behavior. The model would not be greatly altered, however, if producer goods were also included.

8. *All products are ready-made.* Their qualities are therefore chosen, not directly by consumers (as in the case of custom-made goods), but by producers on the basis of estimates of consumers' preferences.

9. *Competition is unfettered.* No restraints are placed on the kinds and qualities of products that firms may offer, or on methods of producing them, or on consumers' selections.

Chapter VI

CONSUMERS' AND PRODUCERS' BEHAVIOR

THE two preceding chapters have been preliminary steps in the undertaking now before us: to show the kind of economic behavior that is induced in a market economy in which consumers, diverse in their tastes and circumstances and uncertain about the actual utility of products, seek to employ their incomes in ways that will maximize the satisfaction of their constellations of basic wants. Through this analysis of behavior patterns it will be seen how demand arises and undergoes modification as part of the economic process of production and exchange.

1. CONSUMERS' BEHAVIOR

First let us look at households. And let us start with the moment at which a consumer becomes conscious of some constellation of basic wants associated with some activity or experience which he feels is worth striving for. Here is the true starting point of economic conduct.

In a rational world any desire to achieve an end invariably implies a desire for the best possible means of accomplishing the end. Ordinarily a basic want cannot be satisfied without the aid of some object or service. The consumer's job, then, is to search for some good or service possessing the qualities that best meet the specifications of his want—or, at least, nearly enough so to make him satisfied not to look further. Here starts what we may call the process of matching wants with goods.

The steps in the process have already been mentioned. First, the consumer must become aware of the existence of one or more products capable of satisfying the basic want (that is, constellation) to some degree. Then he must become acquainted with the performance characteristics of these products—that is, the degree to which they can satisfy each of the various component wants of his constellation. Then, armed with this knowledge, he must mentally weigh the relative advantages of the different products, one against the other, and thus establish some order of preference. Only through these steps does the rational consumer develop wants that become attached to particular products and does desire crystallize into demand.

Charles H. Cooley has described this kind of activity in *Social Process*,[1] terming it "the process of valuation." His homely but clarifying illustration is worth quoting: "Suppose I wish to drive a nail and have no hammer by me. I look at everything within reach with reference to its hammer-value, that is, with reference to its power to meet the special situation, and if the monkey-wrench promises more of this than any other object available . . . it is selected."

It seems likely that individuals, in weighing alternatives, form in their minds more or less consciously some conception of an ideal product: an imaginary product that fits every specification of their want constellation to perfection. This ideal, which I shall term the *optimum conceivable variety*, provides a conceptual target at which to aim when selecting quality; it enables a person to judge the relative merits of competing varieties by comparing each in turn with the ideal and then rating them according to the fewness and smallness of their deviations from it. Some persons, no doubt, have clearer notions than others of what constitutes perfection, depending on their critical abilities and the thought and care they choose to give to the selection process. Some, indeed, may have so rough and vague an idea of "the best" that they may deny altogether employing this method of thinking. Yet it seems reasonable to suppose that consumers either actually do employ this method or else act as if they were employing it.

Having formed his standard of comparison, the rational con-

[1] Charles H. Cooley, *Social Process*, 283–84.

sumer will try to select the *optimum available variety*. This is almost sure to be an imperfect match. For one thing, under our assumption of diversity of wants, the consumer population contains many conflicting notions as to what is ideal; it may even be that no two people have precisely the same set of preferences (just as no two trees are exactly alike). Diseconomies of small-scale production make impractical the production of a sufficient number of varieties to suit the precise needs of every potential buyer. Moreover, the conceptual ideal may be impossible to produce. The ideal durable good is one that will never break, wear out, deteriorate, or require maintenance. The ideal phonograph is one that reproduces the original sounds with perfect fidelity. Products can only approach such an ideal; they can never achieve it. And often a closer approach to the ideal in one respect necessitates a further departure from it in another. H. A. Calahan, an authority on yachting, provides an illustration of this in discussing the wire used in rigging sailboats. Two desirable characteristics in wire rigging are flexibility and freedom from stretchiness. But the properties of wire are such that the more flexible rigging is, the more it will stretch, and vice versa.[2] This is simply one of innumerable instances in which consumers are faced with the necessity of choosing between conflicting ends. In all such cases, the consumer is forced to weigh gain against loss, work out some compromise in his mind, and thus arrive at a modified standard: the *optimum attainable variety*.

Frequently, a person remains wholly unconscious of any discrepancy between the satisfaction sought and the satisfaction gained as the result of a purchase, thinking that the product he has acquired fits his need "exactly." This is especially likely to be so if his wants are not sharply defined in his mind and if he is not a demanding person, or if the horizon of his experience is limited, or if the want is of a sort for which a considerable range of qualities are equally satisfactory, or if the purchase is such a small one that the difference between the closest available match and the perfect match is negligible. Yet actually the possibility always remains

[2] Harold A. Calahan, *Rigging*, pp. 32–36. Calahan lists "eight different kinds of properties which must be taken into consideration in the selection of wire for any specific purpose." This constitutes what we have been calling a constellation of wants.

that some undiscovered or not-yet-produced good might fit his wants even better. This fact becomes most clearly evident when a product represents a fairly complex bundle of utilities. A person who gives careful thought to the purchase of a car or the rental of an apartment normally finds that not one of the possible alternatives suits his taste in every last detail. He would have preferred a composite (which he can easily form in his imagination) of the best features of several of the competing products; but such a choice is not open to him. His final choice is therefore not entirely satisfactory in any absolute sense, although he may correctly consider it to be entirely satisfactory relative to the other available alternatives.

A consumer can choose, of course, only among those alternatives with which he is familiar. Lacking omniscience, he is unlikely to be fully aware of all the possible alternatives. Though a variety already known to him may be highly satisfactory, some undiscovered product might fit his wants even better. Moreover, he may have formed erroneous judgments about varieties known to him. His choice, therefore, will not necessarily be the best product, but what he thinks is the best product. This means that we must distinguish still another kind of "optimum variety": the *estimated optimum:* the variety which the consumer believes to be the best available on the basis of information possessed at the time the estimate is made.

A consumer who is aware that his knowledge is incomplete and possibly faulty will realize that expenditure of effort directed toward acquiring further knowledge may bring worthwhile gains in satisfaction. Consequently an important element in rational consumer activity is acquisition of further knowledge of alternatives through observation, investigation, and experimentation. This means expenditure of time or money or both, on such things as shopping tours, subscriptions to advisory services, study of advertisements, trial purchases of new products, and swapping of experiences with other consumers. No amount of such expenditures, of course, can ever remove completely a consumer's uncertainty as to whether he has discovered the best available product, but since these outlays eventually bring diminishing returns they will be stopped at some point. Whenever this point is reached, the consumer is ready to come to a decision on the basis of the data collected.

The consumer now weighs the alternatives before him. Here we can picture him balancing a superiority in one aspect of a product's quality (as compared with some rival product) against an inferiority in some other respect, in order to determine which outweighs the other in his estimation. At this point in our analysis it will be useful to employ the concept of indifference that forms the backbone of the standard modern analysis of consumer behavior. However, the axes of our *n*-dimensional "diagram" must be conceived as showing, not the amounts of the various commodities a consumer is free to acquire, but the amounts of the various kinds of satisfaction open to him—that is, the components of his constellations of wants. The shape of the indifference function then shows, not the various alternative combinations of products which are equally satisfactory to him, but the various alternative combinations of desired qualities in a product (or set of products) which would make them equally satisfactory to him. An indifference function thus defined reveals the person's underlying value system, which of course is the ultimate determinant of an individual's scale of preferences for commodities.

If we disregard for the time being the complications raised by cost differences and price differences, the solution to problems of choice between competing varieties can be explained in terms of indifference curves used in combination with transformation curves. Let the two axes of a diagram represent two variable properties of a single product: for instance, rigging for a sailboat. Let distances along the *x* axis represent increasing amounts of flexibility, and distances along the *y* axis diminishing degrees of stretchiness (which may be thought of as increasing amounts of resistance to stretching). Starting with some variety of rigging chosen at random, we can plot a transformation function, showing for any given gain in flexibility the decreased amount of resistance to stretching is determined by physical laws and the existing technology.) If rigging were variable in only these two respects, the curve would show the complete set of varieties of rigging producible at that cost. Other curves of a similar character would show the possible pairings of these two qualities in cheaper and more expensive grades of rigging. We can also plot on the same diagram the yachtsman's indifference curves, showing for any given loss of flexibility the needed amount of increase in resistance to stretching in order to make the new combination of qualities just as satisfactory as the old. The optimum attainable variety (among those costing the same to produce) is that which touches the highest indifference curve. The consumer selects that variety among those currently being produced which most closely approximates the attainable optimum.

What can be said about choice between varieties that differ in price? Price is important only in relation to a product's estimated utility. Cheapness of price is advantageous because it leaves the consumer with a larger residual income that can be spent on other forms of satisfaction. Thus alternatives involving price differences can be translated into alternative combinations of utility-yielding qualities, which may then be compared on the basis of one's personal scheme of evaluation. For example, an inferior but cheaper automobile is preferable if the money saved makes possible a television set otherwise unobtainable and if the combination of satisfactions experienced therefrom outweighs the satisfactions that would have been experienced with a more expensive car but no television set.[3]

The variety which provides the best "value for the money" when price differences are taken into consideration is still another kind of optimum. We may call it the *optimum bargain.*

When products differ in both quality and price, the magnitude of the price differential is of crucial importance to consumer choice. In the evaluation scheme of our ideally rational consumer there must exist, for every product which at current prices he regards as

[3] For a similar and more detailed analysis of choice among products which differ in both quality and price see Ruby T. Norris, *The Theory of Consumer's Demand*, pp. 114–27.

the optimum bargain, a critical price (other prices remaining unchanged) above which his allegiance would be transferred to some other variety. Whenever this price is exceeded substitution occurs.

Evaluations of products based on estimates of their satisfactoriness are necessarily provisional. Only when products are put to the test of actual use will the correctness of the estimates be known. In an uncertain world rational consumers will seek to minimize causes for regrets. As errors are discovered, evaluations will be revised and demand schedules altered. Individuals' preferences therefore do not remain constant, being continually modified as discrepancies between their *ex ante* and *ex post* evaluations become apparent. In situations where consumers' tastes and circumstances remain unchanged and no product variation occurs, the discrepancies tend to shrink, perhaps even to disappear. In all other situations the process is a never-ending one, a constant struggle to keep abreast of changing needs and opportunities.

The analysis of consumer choice of quality, it should be noted, can be applied also to business buyers. They too must think of the suitability of products for the particular purposes in mind. The business management of a firm has its own kind of value system, formed as a result of its judgment as to the most efficacious means to maximum profits and/or growth. Certain business activities are considered necessary or desirable. For example, a firm needs to write business letters. This "activity" requires the purchase of typewriters, stationery, and the services of stenographers. It is formally correct but sterile of content to say that the firm will choose that quality of bond paper which will maximize the present value of its expected stream of future profits. Actually, in choosing between various competing brands and grades of stationery the purchasing agent must bear in mind the impression which the firm wishes to make on its customers in order to gain prestige and good will, the efficiency and morale of its stenographic staff, etc., and must weigh in combination such elements as economy, attractive appearance, and ability to withstand erasure. Thus the want for a producer's good also may be regarded as derived from a constellation of basic wants for desirable services in connection with some activity.

2. PRODUCERS' BEHAVIOR

Let us turn now to producers. According to our hypotheses, the data on which their maximizing calculations are based are not the current demands of consumers for the products currently being produced (since these are variables), but rather the public's basic wants, which are associated with certain tastes and preferences but not necessarily attached to any specific goods. To determine the most advantageous product, a producer must first make estimates or conjectures as to the precise character of these basic wants. Then his problem is to determine the quality of product to which these wants will most readily become attached.

If a crude analogy may be permitted, each want constellation of each individual is like a target. Its bull's-eye is the optimum point of satisfaction, provided by the perfect product. Other, slightly different varieties may provide sufficient satisfaction to hit the target at varying distances from the bull's-eye. Since tastes differ, the variety that provides one person with the greatest satisfaction will fall short of perfection in the mind of another; therefore a producer cannot expect a single product to hit, or even come near, every bull's-eye. What he can hope to do, if he wishes to reach a mass market, is to select that product which will maximize his total "score"—that is, hit many targets and come closer to the bull's-eye than his competitors' shots in as many cases as possible. This means, of course, that he cannot ignore the qualities of rival products when making his own selection of quality; his score will depend partly on the pattern of hits that competitors make on those targets at which he is aiming. He may try boldly to outcompete his rivals in reaching the mass market. He may prefer to cater to the wants of a minority whose tastes or needs are sufficiently different from most people's so that their targets are not touched by the products of other firms. Or he may decide to produce several varieties, each tailored to the basic wants of some group of consumers. In any case the pattern that arises is one of product heterogeneity.

Estimates of wants can be based on (1) knowledge of what people have wanted in the past, as expressed in market preferences, (2) investigations as to what people say they prefer, (3) small-scale trial runs of new varieties, and (4) the producer's own ideas, arrived at deductively or intuitively (perhaps partly on the basis of his own preferences), as to what people are likely to want. In many cases empirical study of recent demand is the best single guide to future demand. But the choice of a novel variety hitherto unproduced cannot be made on that basis. Here market research and "pilot" marketing are useful aids. Yet these devices cannot determine which of the numerous possible variants to select for such testing in preference to others. The first-round choosing must be the management's own "guesstimates," based on general familiarity with the public's needs, tastes, activities, and psychology.

Once the product is chosen and production undertaken, the producer is faced with the task so thoroughly explained in economics textbooks of selecting the best output-price combination. But he has also two other important tasks. Since consumers may be imperfectly aware of the existence of his product or its uses, he will find it advantageous to search out potential buyers whose want constellations conform closely to the want-satisfying qualities of his product, and, by informing them, transfer their allegiance from less satisfactory varieties to his own. This means outlays on advertising and sales promotion. And secondly, since his estimates, like those of consumers, are necessarily based on incomplete and perhaps faulty information and cannot be regarded as final judgments, he will also find it advantageous to maintain a skeptical attitude about his product and to search for ways of improving it. Unless the complex of qualities currently embodied in his product happens to be the ideal combination (a remote possibility), quality changes of the right sort will improve his position in the market. Thus the incentive to bring out new models and to create new products.

In performing these tasks every producer is working more or less in the dark. He is in the position of the blindfolded person who is trying to pin the tail on the donkey, and whose only way of judging his degree of success is by listening for sounds of approval or titters of amusement from the rest of the gathering. Conjecture, imagination, venturesomeness, research and investigation, trial and error, sensitiveness to consumers' reactions—all these are of the essence of entrepreneurial behavior.

According to this picture product variation is the experimental process by which producers test new hypotheses concerning consumers' basic wants and the qualities of goods which will best serve these wants. Since absolute certainty is unattainable, it is a never-ending process.

Provision of information to consumers is likewise a never-ending process. Individuals forget, and need frequent reminders. Progress and change aggravate the need for it; but even in a stationary society young people would mature and climb to higher rungs of the economic ladder, with the result that there would be continual shifting of individual members of the population into and out of markets, so that fresh information would constantly be required.

Actually, a stationary state could not exist under the conditions we have assumed. Our analysis has revealed an evolutionary process that is inherent in a capitalist free-market society, that *must* tend to go on so long as producers lack omniscience, possess some degree of ingenuity, and want to expand sales. Here our model conforms well to what we can observe in the real world. New products and new variants come into existence. Old products grow or wane in popularity, or pass into oblivion. Small innovations are persistently being sought and carried out even in conservative industries and firms in an effort not to fall behind in the competitive race. A substantial part of managerial effort consists of continual playing with new ideas. This ever-present force operating to produce evolutionary change is more noticeable in some market areas than others; in the case of a few staple commodities the process remains wholly dormant. Yet in general one of the most striking characteristics of competitive markets is their evolutionary character.

This is of course the very same process that Schumpeter has analyzed so well in his studies of innovation. Schumpeter purposely devoted his attention mainly to the large, revolutionary innovations, such as the power loom, the railroad, the automobile. He conceived innovation as a "change . . . of the first . . . order of magnitude," and defined "development" to exclude change that takes place through "continuous adjustment in small steps." [4] Thus he was able to describe innovation as an intermittent, discontinuous process, which operates "by fits and starts"; the head of a firm, he declared, ceases to be an innovator "as soon as he has built up his business, when he settles down to running it as other people run their businesses." [5] Yet actually no such sharp line can be drawn between grandiose innovational activity and the humbler changes which characterize the intelligent running of an established business. Schumpeter himself was careful to say that "we must try to divest ourselves of the idea that innovation necessarily means something spectacularly important." [6] Nevertheless Schumpeter has pictured innovation as a spontaneous activity on the part of a few men of great imagination and daring; and Paul M. Sweezy has rightly criticized this picture on the grounds that it overlooks the constant pressure of modern competitive conditions that forces business firms to keep innovating in order to preserve profits. "Nowadays," writes Sweezy, ". . . the appearance of important new firms is a rare event. . . . Innovation is carried out largely by existing enterprises almost as a part of their regular routine." [7] If product innovation is conceived to include the minor as well as the major revisions in products, it is seen to be, under the conditions assumed, not an occasional interruption of the routine process of exchange but an integral part of it.

It is important to note that according to the line of analysis being presented here, innovation is an equilibrating process, even though it is of course in the immediate and usual sense a disequilibrating element, upsetting cost and demand schedules. It is equilibrating in that it represents a striving toward that never-to-be-reached equilibrium position in which all products possess the qualities which, within the limitations of nature's laws and the other unalterable elements in man's environment, best serve consumers' basic wants.

[4] Joseph A. Schumpeter, *Business Cycles*, Vol. I, p. 94; *The Theory of Economic Development*, pp. 65–66.

[5] Schumpeter, *The Theory of Economic Development*, pp. 62, 75, 78, 81.

[6] Schumpeter, *Business Cycles*, Vol. I, p. 92.

[7] Paul H. Sweezy, "Professor Schumpeter's Theory of Innovation," *Review of Economic Statistics*, XXV (Feb., 1943), 96.

Innovational changes in products, moreover, like price changes, are often equilibrating *responses to disturbances,* and form a part of maximizing behavior. A change in consumers' circumstances or in the technical knowledge available to producers may alter the set of qualities which would be judged optimum by consumers of given tastes. (A change in the quality of roads, for instance, may affect motorists' scales of preference for various types of spring construction in automobiles.) If so, those consumers would be better off if a new variety were produced embodying the nearest possible approach to the new set of optimum qualities. Any producer who wished to have these consumers as his customers would improve his position if he altered his product to conform to the new set of qualities. Whatever change occurred as a consequence would be a movement toward a new equilibrium.

3. MODIFICATIONS IN THE ASSUMPTIONS

Thus far our picture of producers' behavior has been one of a world in which producers strive earnestly to select those qualities that best fit the underlying tastes and aspirations of consumers. Their business success hinges on their success in doing so, which in turn depends on the degree to which they are able to penetrate the fog of partial ignorance concerning consumers' wants and possible ways of redesigning products to make them serve such wants better. In a world in which consumers were perfectly rational and producers perfectly ethical, no other methods of influencing demand could provide a path to growth and profits.

Now let us relax these two assumptions. If some consumers are sometimes gullible or thoughtless or lazy enough to accept producers' claims uncritically, and if some producers are willing to exploit those consumers, new forms of behavior are introduced into the model. Some producers will succeed in creating new demand by making false or misleading claims and by concentrating on qualities and gadgets that make a product look better than it is, while skimping quality wherever skimping cannot be easily detected. They will also find it profitable, in instances where betterments cannot be thought up, or are too costly to be worth while, to introduce quality differences that are actually irrelevant to consumers' wants yet can be successfully advertised as "improvements" and "exclusive features." Their efforts will thus be directed partly toward increasing consumers' ignorance, or at least perpetuating it. Rational consumers must now be on the lookout for chicanery as well as honest error in appraising producers' claims; their task will thus be made more difficult. In these instances producers and consumers will be working at cross-purposes, and some canceling out of effort will result. Some consumers will persistently buy some products under the illusion that they possess qualities which in fact they do not possess.

But unless all consumers are wholly undiscerning and unintelligent and all producers wholly rapacious and lacking in creative ambition, the new forms of behavior described above will not replace the forms described in the preceding sections but will be superimposed on them. It should be noted that the new forms of behavior are not the result of uncertainty alone but of the combination of uncertainty, irrationality, and low standards of ethical conduct. If any one of these conditions were removed, persistent exploitation of ignorance could not occur.

Let us now restore the assumptions of perfectly rational, ethical behavior, and assume in addition that producers have perfect knowledge (while consumers, as before, remain imperfectly informed). This is a useful modification, since in fact the typical business firm, being a specialist, is far better informed about products' qualities and the general pattern of consumers' activities, needs, and tastes than is the typical consumer. If producers' knowledge is perfect, experimental product variation becomes unnecessary. And if the structure of basic wants of the population as a whole remains unchanged, producers will at once select those varieties that will maximize consumer satisfaction, whereupon product variation will cease entirely. So long as consumers differ in their tastes and environments their optimums will differ, and numerous varieties of each type of product will be marketed. And so long as consumers' knowledge is imperfect, marketing activity—investigation and recourse to expert opinion on the part of consumers, and informative advertising and sales promotion on the part of producers—will also be found. If we now relax the assumption of unchanging tastes and values, product variation reappears, as qualities are altered to conform to the new optimums. Possibly we should expect to find fewer changes than would occur under producer uncertainty.

If consumers as well as producers had perfect knowledge, investigation and advertising would no longer be needed, and would disappear. Competing products would still be heterogeneous, however. And product variation, similarly, would arise only when changes occurred in people's constellations of basic wants.

Now let us return to the original model with the single change that all consumers are assumed to have identical tastes, environments, and incomes, and therefore identical basic wants. Under these circumstances consumers would have identical scales of preferences for the possible variants of any product, and a single conceptual optimum variety of each type of product would be shared by all. The only basis for differences in individual demand schedules would be errors or gaps in knowledge. These would tend to be corrected more quickly than in a world of dissimilar basic wants, since everyone would gain from everyone else's experience. Entrepreneurs, being also consumers (and therefore, by hypothesis, having the same basic wants as their customers), would need only consult their own preferences to determine whether their own or some rival product were better adapted to consumers' wants. Thus inferior varieties would be weeded out, and differences in rival products could not persist. Of course occasional quality changes would be made, due to entrepreneurial ingenuity in discovering improved ways of satisfying existing wants, but each new improvement would in due time become standard. Whether the predominant characteristic of markets would be homogeneity or heterogeneity of product would depend on the frequency of innovational quality changes and the length of time required for their adoption by all consumers and producers. If we relax the assumption of constant basic wants, a new factor making for product variation is introduced. These new conditions, together with lags in the communication of knowledge and in the technical adjustments required in order to get the new varieties into production, might be sufficient to make homogeneity of product as rare as it is in the actual world.

To assume a world in which all people have identical basic wants is perhaps unduly artificial, yet it is instructive in that it reveals the thoroughly competitive character of some kinds of quality variation. When all buyers are alike, certainly quality adjustments cannot be interpreted as attempts to secure a sheltered market by appealing to the special tastes of some subgroup of buyers; they are, on the contrary, means by which a producer tries to serve with superior skill or appeal the needs of the same group of buyers that his rivals are attempting to serve.

Suppose we return to the original model with the single change that individuals are now assumed to be unable to improve the state of their knowledge. This single alteration is sufficient to effect a complete transformation in consumers' and producers' behavior. In a world in which consumers can never learn anything new, investigation, experimentation, and informative advertising are bound to be futile and will be abandoned. Nor will product variation be of any avail. Derived wants will remain as fixed as basic wants. Our model now conforms to the stationary state of the textbooks.

Again we return to the original model, this time with the single change that consumers' tastes and values are assumed to be variable, and subject to influence. Now producers will find it possible to persuade some consumers to acquire the tastes and basic wants which their products can best satisfy. They will not be able to do this at will, since consumers are rational; but it is not irrational for a person to develop new tastes, and when he is in the process of doing so an outsider can often accelerate the process or even guide it in a slightly new direction by prodding his imagination or by persuading him to try some new experience. Perhaps the easiest way of changing a person's basic wants is to induce him to undertake some new activity, such as fishing, photography, or sightseeing. (This does not necessarily involve a change in underlying tastes, though it may. It could consist merely of informing a person about an activity which he would have liked to have undertaken all along if he had been fully aware of its merits.) It is also possible to persuade people to develop a preference for carrying on old activities in new ways. Under these conditions producers will devote some part of their advertising and sales efforts to endeavors to change consumers' tastes, while consumers will also devote some of their efforts to seeking new wants and tastes through investigation and exploratory activity. It is impossible to say whether or not the volume of these efforts will be substantial. Offhand, one may guess that producers will generally find it easier and less costly to gain sales by adapting the product as closely as possible to existing tastes and by directing advertising to those whose wants it is already well equipped to satisfy than by attempting to alter human beings to fit the product. If such is the case the modification will be only of minor importance.

4. A SUMMARY VIEW

What we have done so far is to sketch out rather roughly a theory of the consumer and a theory of the firm which explain economic choice among alternative qualities as a process of adapting means (product quality) to ends (constellations of basic wants). By exploring the implications of two assumptions not usually made explicit in formal theory—namely, heterogeneity of wants and uncertainty about utility—and by focusing our attention on matters other than costs, prices, and outputs, we have obtained a view of the exchange process that has disclosed driving forces of economic conduct in competitive markets which have heretofore been largely neglected in formal economic theory.

The kinds of behavior described in this chapter are of course familiar to everyone; in fact, the reader may have an uneasy feeling that nothing new has been added to his understanding of economic activity. What novelty this chapter possesses lies, not in the behavior patterns discussed, but in the revealing of their relationship to fundamental wants and environmental conditions. They are shown to be, not simply "fringe" activities, but part and parcel of the basic process of economizing.

The following statements may now be made:

A. *Propositions Concerning Consumers' Behavior*

1. Consumers' basic wants are wants for activities that provide satisfying experiences. The detailed specifications of the kinds and amounts of satisfactions sought through these desired activities constitute the ends of consumers' economic conduct. A "want" for an activity or experience is not a single want but a constellation of wants.

2. From these constellations of basic wants are derived wants for whatever material objects and personal services are thought to be of greatest aid in achieving the desired activities.

3. A person's basic wants are a function of his inherent tastes and values, and of his circumstances (occupation, income, size of family, place of residence). His circumstances are partly a function of his inherent tastes and values, but not wholly so.

4. A person's derived wants are a function of his basic wants and of the extent of his information (or misinformation) about the various alternative means of satisfying his basic wants.

5. Rational consumers whose ends are fixed remain open-minded as to the most appropriate means to those ends. Being never absolutely certain that they are fully and correctly informed, they consider their estimates of products and thus their scales of preference for them to be provisional judgments, subject to revision whenever new information or experience reveals errors.

6. Derived wants are changeable, even when basic wants remain constant. A derived want will disengage itself from one product and attach itself to another whenever a person believes that the new product is better capable of satisfying the basic wants than the old.

7. Corresponding to every constellation of basic wants of an individual there must be an optimum attainable variety of a product —that is, that variety which, if produced, would provide the maximum satisfaction of the wants. If an optimum variety could actually be bought by an individual, no further gain in satisfaction could be realized by shifting to some other close substitute. An optimum variety (or an optimum bargain, if price differences are a consideration) represents the final equilibrium position of a derived want.

8. Movements toward this equilibrium are effected by improvements in knowledge. By this process the estimated optimum approaches the actual optimum, and in equilibrium becomes identical with it.

9. Deliberate investigation and experimentation to improve knowledge are necessary elements in rational consumer behavior.

10. Acquisition of knowledge generally involves costs (monetary or nonmonetary) which must eventually bring diminishing returns. When this is the case, differences between consumers' estimated optimums and their actual optimums may be expected to persist indefinitely.

11. Rational choosing requires evaluation of the trustworthiness of sources of information. Sources of repeated misinformation will tend to be discredited.

B. *Propositions Concerning Producers' Behavior*

1. Producers desire to maximize the demand for their products. Therefore they seek to attach consumers' derived wants to their own products.

2. A producer has two possible methods of creating a demand for his product (at any given price) on the part of a consumer who has refrained from buying: (*a*) he can alter the consumer's attitude toward the product as it stands, or (*b*) he can modify the product so that it conforms more closely to the consumer's existing preferences or prejudices.

3. Consumers' attitudes toward existing products can be altered by (*a*) improving the extent or accuracy of their knowledge of the products and their uses, (*b*) creating erroneous beliefs favorable to the product, (*c*) inducing unreasoned acquiescence to simple suggestion, or (*d*) altering consumers' tastes and values, and hence their basic wants. All four methods involve the use of advertising or other forms of sales promotion. Under conditions of perfectly rational, ethical behavior and constant tastes, only the first method is possible. When these conditions are relaxed the first method is not abandoned—it continues to be advantageous for some producers or for some elements of a product's quality—but is employed in conjunction with the others.

4. The original selection of the product and its later modifications are made on the basis of estimates of consumers' basic wants and tastes. Producers do not have exact knowledge and so can never be absolutely certain as to the accuracy of their estimates. They must therefore consider them to be provisional judgments, subject to revision whenever new evidence reveals errors.

5. Investigation of consumers' activities, habits, and tastes is an important element in rational producers' behavior.

6. Products are variable even when basic wants remain constant. A producer will alter the qualities of his product whenever he has reason to believe the new qualities better capable of satisfying consumers' basic wants than the old.

7. Variations in basic wants provide a further reason for product variation.

8. The qualities which a producer may most advantageously select are a function of the current qualities of his competitors' products. The wants that are most readily served are those that have been most neglected by other producers.

9. Producers will not seek to copy rivals' products exactly but rather to adopt some unique details in quality that may be judged superior by at least some consumers.

10. Product variation is the experimental process by which producers test new hypotheses concerning consumers' basic wants and the qualities that will best serve these wants.

11. Innovational changes in consumers' goods are not occasional spontaneous interruptions of the routine of maximizing conduct but are an integral part of it.

C. *Propositions Concerning the Characteristics of Competitive Markets*

1. The number of optimum varieties of any type of product in a community depends on the number of dissimilar constellations of wants in the community which products of that type are capable of serving. In a community of *n* persons, the number will be less than *n* only if different persons have constellations which are identical, or virtually so.

2. So long as tastes and environments differ, constellations of

wants will differ, and no single product can be the optimum variety for a whole community.

3. In any kind of economic system a market of differentiated products can cater more precisely to individual wants and supply a greater sum total of satisfactions than a market devoted to a single homogeneous product. Unless production of more than one variety involves increases in costs so great that they more than offset the higher satisfactions, a single standardized product without close substitutes cannot maximize welfare.

4. With the exception just mentioned, a situation in which a single standardized variety is the only one available cannot be a position of equilibrium. Some producers could improve their positions by altering product quality, and some consumers could improve theirs by purchasing the new varieties.

5. Qualities of competing products that serve the same general purpose do not ordinarily tend to become identical. In a freely competitive economy, in which design and quality of products are not restricted, competing producers will normally vary their products one from another, each either seeking to appeal to tastes and preferences neglected by his competitors or believing that his own choice of variety conforms most closely to the majority's wants. The result will be a market of differentiated products.

6. "Monopolistic competition" (that is, competition between dissimilar products serving the same purpose) is the normal type of competitive market structure. Only under special, rare circumstances does free competition lead to "pure competition" (that is, in which the competing producers are offering identical products). This would be true even in a Utopian society.

7. Uncertainty is ineradicable from economic life, but can be reduced by investigation, experimentation, and improvements in the communication of knowledge. An important part of economic activity is concerned with learning, and with making use of newfound knowledge.

8. Intelligent decision-making requires the aid of institutions whose purpose is to make knowledge more perfect, and in a rational society these tend to arise. In this category are consumers' testing

and advisory organizations, advertising, market research, and experimental product variation.

9. The nature of the competitive process forces entrepreneurs to take an active part in creating change—and hence in *creating* uncertainty as well as responding to it. This augments the need for devices to reduce it.

10. Creation of demand is as essential a part of market processes as is the supplying of demand. Expenditures on advertising and investigation are a normal concomitant of competitive exchange.

11. The qualities of products tend to vary through time, due to forces inherent in the competitive exchange process.

12. Quality changes tend to breed further changes. A variation in the product of one firm will upset the product equilibrium for all the others.

13. Evolutionary change is part and parcel of the exchange process, an inevitable consequence of maximizing behavior.

The above conclusions may be drawn independently of the presence or absence of consumer irrationality, deceptive sales techniques, technological change, or the variability of consumers' underlying tastes. Nor do they depend on the depth of buyers' ignorance or the difficulty of remedying it. Improvements in knowledge may alter the character or scope of product differentiation and product variation, but they do not tend to eliminate or inactivate or even necessarily diminish these phenomena. The determinants of the extent of product differentiation and variation are rather such things as the degree to which it is technically possible to vary the product (which depends partly on the complexity of the product), the variety of the uses to which it can be put, the cost of adequate quality comparison, and the importance to the consumer of satisfying his constellation of wants precisely rather than roughly. Nor do the conclusions depend on the assumption that producers seek to maximize profits; the motive to excel, or to acquire power, or to "win the game," or to engage in empire building would be sufficient to explain the kind of producers' behavior and resultant market characteristics sketched above.

Chapter VII

SOME DISTINCTIONS MADE POSSIBLE BY THE THEORY

THE theory developed in the previous three chapters makes possible several distinctions which cannot be drawn from the current theory of competitive markets, yet which are useful in economic analysis, especially in questions involving welfare and policy.

1. THE BOUNDARIES OF AN INDUSTRY

It is often useful to distinguish commodities belonging to the same industry or group from those which lie outside the group. The very concept of oligopoly, share of the market, dominant firm, and ease of entry and the meaningfulness of studies of concentration and oligopoly in present-day economic life, depend upon such a distinction. A substantial part of the theory of monopolistic competition cannot be stated except in terms of some definable group of "rival products."

Yet attempts to define an "industry" or group unambiguously have not been successful. It is easy enough, of course, to agree on the boundaries of an industry operating under conditions of pure competition, when competing products are perfect substitutes; but when substitutes are imperfect, there seems to be no exact point at which to draw the line.

The time-honored practice of grouping together firms making products which bear some generic name (for example, "motor vehicles") has been rightly criticized on the grounds that it can lead to one's classifying products as "competing" that are hardly substitutes in any sense of the word (for example, sports cars and heavy-duty trucks). For some purposes, of course, it is useful to group business firms into broad categories—electrical products, textiles, steel, pharmaceuticals—and label them as industries. The term "industry" is rather widely employed in this sense. But for the study of market structures, and of questions concerning the degree of control over price and the availability of alternative sources of supply, these categories are far too broad; they furnish no measure of the competitiveness of markets in which these products are sold. When dealing with such questions the necessity for a narrower concept of industry is well recognized. Here it is logical to classify products as belonging to the same industry if they are "close" substitutes.

But the term "close substitute" turns out to be painfully elusive. For contemporary theory holds that substitution is definable only in terms of cross-elasticity of demand; "closeness" is therefore continuously variable, and any dividing line between "close" and "remote" is arbitrary.[1] Joan Robinson has suggested, as a practical expedient, grouping products that are separated from others in the general chain of substitutes by distinct gaps; but this does not remove the arbitrariness of the dividing line, for the size of a gap also is a continuous variable. Mrs. Robinson herself admitted that "there must be some arbitrary element in drawing the boundary, and all products must be regarded as a continuous series in more or less close rivalry with each other."[2]

The result has been that many economists, following Triffin and Chamberlin,[3] now regretfully conclude that the concept of "industry" is valueless for purposes of economic analysis. Thus, in a highly important area of economic inquiry, theory has failed to provide much-needed help, and economists have been forced to fall back on improvised yardsticks.

1 See George J. Stigler, *Theory of Price*, 1st ed., p. 239.

2 Joan Robinson, "What Is Perfect Competition?" *Quarterly Journal of Economics*, XLIX (Nov., 1934), p. 114.

3 Robert Triffin, *Monopolistic Competition and General Equilibrium Theory*, pp. 86, 89, 140; Edward H. Chamberlin, "Product Heterogeneity and Public Policy," *American Economic Review: Papers and Proceedings*, XL (May, 1950), 86–87.

If the concept of basic wants is adopted, however, a criterion for defining the market or industry is provided. A group of heterogeneous products can be viewed as a set of variable means to the satisfaction of some common set of ends—those contained in some specified constellation of basic wants. A constellation is related to some planned activity, which may call for a single physical object (for example, a magazine) or may require or permit the simultaneous or sequential use of several complementary products (for example, camera, film, developing outfit, projector). In either case the constellation of wants in the consumer's mind when he enters a market consists of wants for the services which a single product is capable of rendering. (If he lacks more than one of a group of complementary products, it is best to think of him as having more than one constellation.) To any such consumer the alternatives to be evaluated and compared are the various available products capable of satisfying the constellation. Products which can serve the wants in the constellation belong within the group; products which cannot possibly serve them lie outside the group.

The industry, then, consists of all those products, any one of which is capable of making a particular type of contribution toward the attainment of some experience or activity. When several products are used jointly in some activity, all products capable of making one type of contribution form one group or industry, while those capable of making a different, complementary type of contribution belong in another industry. Thus razors, blades, shaving brushes, and shaving creams belong to four different industries, while electric shavers, safety razors, and straight razors belong to the same industry.[4]

Another, less technical way of explaining the industry concept just outlined is by employing the idea of "purpose" or "function." A product ordinarily serves some purpose, or has some function, that can be described in simple language. The function of a razor

[4] Since the term *industry* is often used with a broader meaning, it may be preferable to employ another term to indicate the group of producers or products that compete in the sense of offering alternative means to the satisfaction of a given constellation of wants. If so, I suggest the term *field of competition*, or simply *field*, to designate the group thus defined. I have thought it best, however, to refrain from burdening readers with a newly coined term, since *industry* is commonly used in economic theory in this sense.

is to remove whiskers; the function of toilet soap is to cleanse the skin; lawn mowers serve the purpose of cutting grass short. A collection of products that perform the same function may be said to belong to a single industry.

According to this analysis, the relationship between products that are capable of serving a common purpose is different in kind from the relationship between products that serve different purposes.

Producers' goods too can be grouped into industries by this principle. Business firms engage in activities which require acquisition of products. Builders, for instance, engage in home construction. They need space-heating equipment in order to make the homes acceptable to buyers. Products capable of serving that want belong to the space-heating industry, from which other kinds of building equipment (water heaters, for example) are excluded.

In all likelihood the products within an industry as here defined would correspond almost exactly to those selected on the basis of cross-elasticity. Even so, the concept proposed in this section seems superior on two counts. It explains *why* some products are interdependent to a significant degree, and others not. And it provides a less laborious method of determining the boundaries of an industry than that of calculating innumerable cross-elasticities.

Industry boundaries cannot be marked off with absolute precision, of course, since buyers do not have identical constellations of wants. Borderline products exist which some buyers might rate as minimally acceptable (provided nothing more suitable showed up) but which others might consider too inappropriate to use under any circumstances. Some people, for example, prefer to forego automobile transportation altogether rather than drive an ancient, shabby vehicle; others are less demanding. Again, some people may be prospective buyers of diamonds (and will accept no substitute), others of precious stones (including diamonds), others of jewelry (not necessarily precious stones). There is no rule by which to decide which of these concentric circles is the most appropriate boundary.

This is not a fatal objection, however. Most scientific classifications (for example, animal vs. vegetable matter) have boundaries that are similarly blurred. Industry boundaries cannot be expanded or contracted indefinitely. And they can usually be set within fairly narrow limits by the conventions of modern society, which prescribe wanted activities and experiences in rather specific terms. In applying the distinction to concrete cases, any unavoidable ambiguity of boundaries is likely to be a much smaller handicap to the analyst than the arbitrary character of boundaries established by selecting some numerical value of cross-elasticity. Moreover, once an industry's boundaries have been roughly determined according to the concepts outlined above, it may then be both practical and expedient to determine whether to include or exclude certain borderline products on the basis of their cross-elasticities.

Another complication must be recognized. Differences in consumers' incomes and also in geographical location divide an industry or market, as defined above, into overlapping submarkets. Few persons debate whether to buy a Henry J. or an imported Daimler, or whether to do their week-end supermarketing in Bridgeport or Trenton. These differences complicate the analysis of market structure by adding other factors which need to be examined, and which may further narrow the actual market area. Yet they do not invalidate the basic distinction between products which can, and those which cannot, serve a defined purpose.

The distinction drawn is essentially a distinction between two different kinds of substitution: (1) intra-industry substitution, in which one means of satisfying a constellation of wants is substituted for another, and (2) inter-industry substitution, in which one constellation is left unsatisfied, or less completely satisfied, while the funds previously devoted to it are employed in satisfying another. The elimination of one's favorite brand would necessitate the first kind of substitution; a complete shutdown of an industry's production would necessitate the second kind.

If this distinction is acceptable, the concept of "freedom of entry" becomes meaningful and useful again. This seems a decided gain. To hold, as Chamberlin does, that "freedom of entry" is meaningless since no one can be free to duplicate another's product in all respects, while everyone is free to produce some kind of product, is like insisting that "freedom of location" is meaningless in connection with watching a parade since no law-abiding citizen is free to stand in a spot already occupied by someone else, and no police regulations, no matter how tyrannical, can prevent a person from being in *some* location in this world. After all, if one's goal is to watch a parade, it does make a difference whether one is permitted to stand within sight of it or is forced to remain several blocks away.

2. THE TWO KINDS OF MONOPOLY

If the distinction between the two kinds of substitution is valid, it becomes possible to distinguish two kinds of monopoly. One kind exists if a given variety of product is produced by only one firm. In this event consumers can either assent to the monopolist's terms or turn to some substitute means of satisfying the constellation of wants being served by that product. On the other hand, a monopoly may embrace an entire industry. In this case the power possessed by the monopolist is of a different kind; control is now held over all avenues to the attainment of the desired experience; consumers must either assent to the monopolist's terms or forego the experience. This second kind can be termed *industry-wide monopoly.*

The distinction seems a useful one. Even if the disadvantage of monopoly is conceived to lie solely in the slope (that is, less than perfect elasticity) of the seller's demand curve, which enables him to exact a higher price, there is reason to expect a much greater slope (lower elasticity) in the case of an industry-wide monopoly than in the case of a monopoly of only one of several avenues to the satisfaction of a constellation of wants. If so, the divergence of the monopoly price from the perfectly competitive price is likely to be far more substantial in the case of industry-wide monopoly. But there is another difference between the two kinds of monopoly. The industry-wide monopolist has sole power to determine what alternative qualities shall be made available. As a consequence the alternatives offered may be too few, or the wrong ones. And if no other producer is free to enter the field, the monopolist is not under the same pressure to re-examine the appropriateness of his products to the wants of his customers as is a firm which has merely a monopoly of one of numerous varieties. Even a firm having the best of intentions for serving its customers in the best possible way may fail to perceive opportunities for product improvement that would be perceived by at least one of a number of independent producers in competition with each other.

Industry-wide monopoly as defined above corresponds to the concept of monopoly generally employed in economic theory prior to 1930—for example, Ely's definition of "monopoly in the economic sense" as "giving exclusive control of one sort of business." [5] It also corresponds to the concept of monopoly which the lawmakers who contrived the antitrust laws and the courts which have dealt with alleged violations have generally had in mind.[6] This is not to say that the existence of the other kind of monopoly was not recognized, but "a monopoly in a specific article" was held not to be harmful unless it led to "a monopoly in any branch of trade." [7]

Single-variety monopoly, on the other hand, is the kind chiefly emphasized in modern price theory. Wherever products or their conditions of sale are differentiated, each seller is viewed as a monopolist of his own product. Moreover, with the breakdown of the industry concept, the belief has arisen that the two kinds of monopoly are not different in kind but only in degree. It is pointed out that every seller of a differentiated product, though possessing "an absolute monopoly of his own product," must face "the competition of more or less imperfect substitutes." But "since all things are more or less imperfect substitutes for each other" precisely the same thing can be said of the most powerful trust or cartel. "The difference between the Standard Oil Company in its prime and the little corner grocery is quantitative rather than qualitative." [8] Ac-

[5] Richard T. Ely and others, *Outlines of Economics*, 5th ed., 1930, p. 562. By contrast, monopoly "in the strictly legal sense" includes trademarks and patents.

[6] See Edward S. Mason, "Monopoly in Law and Economics," *Yale Law Journal*, Vol. 47 (Nov., 1937), 34–49.

[7] See, e.g., the excerpt from the report of the Committee on Patents in the House of Representatives in 1912, quoted in Chamberlin, *Monopolistic Competition*, p. 58.

[8] Chamberlin, *Monopolistic Competition*, pp. 9, 65; Harold Hotelling, "Stability in Competition," *Economic Journal*, XXXIX (March, 1929), 44. Not all contemporary writers take this position, however. Some still find it fruitful to treat industry-wide

cordingly, monopoly is defined in the standard formulations of modern price theory as any situation in which a single seller faces a less than perfectly elastic demand.

This definition has obvious disadvantages. If it is almost universally true that competing goods are not identical in all respects, the word "monopolist" becomes practically synonymous with "seller" and "competitor." And if, in general, it is normal and desirable that this should be so, elimination of monopoly in this all-inclusive sense not only is impossible but would be inimical to welfare. The word "monopoly" loses its sharpness and usefulness as a tool of analysis.[9]

The modern theorist might answer this criticism by saying that this is no disadvantage; if the older distinction is in fact artificial and false, it is an advantage to realize that in a capitalist economy monopoly is everywhere and ineradicable. Yet the stubborn fact remains that monopoly takes on a different aspect as a social problem when it approaches or reaches industry-wide proportions. One qualitative difference which even contemporary analysis brings out is that complete or partial monopolization of an entire industry results in there being too few rivals whereas single-variety monopolization results in there being, if anything, too many. Another, pointed out by Triffin, has to do with the stability of the firm's demand curve: "The real significance of the traditional monopoly case—the singleness of the seller—is to identify a firm with an industry and thus to substitute for the shifting demand curve of the former the stable demand curve of the latter. It is in this sense that monopoly means the absence of competition." [10] And many practical economists remain convinced that, formal theory aside, public

---

monopoly as something qualitatively different from brand monopoly. See, e.g., John Maurice Clark, *Alternative to Serfdom*, p. 64; J. K. Galbraith, "Monopoly Power and Price Rigidities," *Quarterly Journal of Economics*, L (May, 1936), 458; and Vernon A. Mund, "Monopolistic Competition Theory and Public Price Policy," *American Economic Review*, XXXII (Dec., 1942), pp. 737–39. But their viewpoint seems to be the exception rather than the rule.

[9] Cf. C. E. Ayres, *The Theory of Economic Progress*, p. 184: "Words . . . are tools. . . . But if it is to be a tool, it must cut. It must distinguish something or other from something else. To apply the word . . . to every sort of activity and behavior function is to destroy its cutting edge altogether, and thereby to reduce its use to that of a much coarser word-tool . . . with which we are already adequately equipped."

[10] Triffin, *Monopolistic Competition and General Equilibrium Theory*, p. 131.

welfare is not seriously menaced unless monopoly extends beyond control over the supply of a single branded good.[11]

Chamberlin, conceiving all monopolies as alike in kind, identifies monopoly with product heterogeneity. And since he finds that heterogeneity is part of the welfare ideal, he is forced to conclude that monopoly also is necessarily a part of the welfare ideal, and that the road to higher welfare may sometimes involve making the economy "more monopolistic" rather than "more competitive." [12] If, however, the distinction between the two kinds of monopoly is employed, a different and perhaps more useful framework of analysis can be built. Product heterogeneity can be defined as the dissimilarity of competing substitutes (that is, goods that can satisfy the same constellation of wants). Single-variety monopoly is normally a concomitant of product heterogeneity, but industry-wide monopoly is not—in fact it may be an obstacle to sufficient heterogeneity, and in the case of a single-product firm it prevents heterogeneity altogether. If we accept the proposition that product heterogeneity is a part of the welfare ideal, our ideal will include some degree of single-variety monopoly. But it might well be that industry-wide monopoly is always inimical to welfare. Here at least is a hypothesis to be investigated which cannot even be framed when the blunter tools of the Chamberlinian analysis are used.

3. THE TWO KINDS OF WANTS

An important distinction already made, which needs only be mentioned at this point, is the distinction between wants for experiences (basic wants) and wants for commodities (derived wants). At present a good deal of ambiguity is involved in statements that "consumers do not know what they want," or that "wants are unstable," or that "the purpose of most advertising is to change people's wants." Such statements may be true of one kind of want while being untrue of the other kind. And whether it is intended to refer to one or the

[11] See, e.g., Edward S. Mason, "Various Views on the Monopoly Problem," *Review of Economics and Statistics*, XXXI (May, 1949), 104, 105–6.

[12] Edward H. Chamberlin, "Product Heterogeneity and Public Policy," *American Economic Review: Papers and Proceedings*, XL (May, 1950), 86.

other is not unimportant, for there is a world of difference between changes in consumers' goals and aspirations and changes in their day-to-day thinking as to the best probable way of achieving these goals. If the two can be distinguished, greater precision of statement is possible.

4. ACTUAL VERSUS OPTIMAL

If ends are indicated by a person's basic wants, and selection of means by his derived wants, it may be said that a person's derived wants are optimal whenever the products wanted are the best possible means to the ends currently being sought. This gives us a criterion of an individual's economic welfare. The extent to which an individual's actual derived wants differ from his optimal derived wants is a measure of the extent to which he has fallen short of achieving maximum welfare.

Consumers may choose incorrectly because they are misinformed, or have misplaced faith in others' recommendations, or have somehow acquired illusions which they irrationally retain, or have been frightened or cajoled or hypnotized into buying what they would not otherwise have bought. No matter what the cause, they are receiving less than maximum satisfaction from their outlays, even though they may never realize it.

Contemporary economics is aware of the prevalence of faulty choosing, yet it possesses no precise conceptual tools for determining whether or not the products selected are optimal. The recent literature, it is true, abounds with contrasts between what people "ought to" want and what they "actually" want, between what they "really" want and what they "are made to" want, between the "objective" and "subjective" elements in utility and demand, between "rational" and "irrational" preferences, between "actual," "real," "genuine" differences on the one hand and "fancied," "illusory," "supposed," "nominal," or "pseudo" differences on the other. The trouble comes when an attempt is made to define these terms.

First it should be noted that some theorists hold that it is impossible to distinguish between the imaginary and the real, since utility and satisfaction are in any event subjective matters. Knight presented this viewpoint in *Risk, Uncertainty, and Profit.* According to his analysis "an element of appeal due to a high-sounding name" is not essentially different from "elements in the physical form and appearance of a commodity" since "these things do make a difference in the commodity to the consumer and in an exchange system the consumer is the last court of appeal. If they are different to him, they are different. . . . I do not see that it makes any real difference whether these utilities are in the thing itself or in some associated fact." [13] Chamberlin, following this line of reasoning, has argued that "when one producer copies the name, symbol, package, or product of another, the result is goods more nearly standardized," thus implying that there is no significant economic difference between misrepresentation of origin and imitation of another product's qualities.[14]

There are however two serious drawbacks to the wholly subjective type of economic analysis. One is that any contribution to welfare economics is impossible—unless it is claimed that it does not matter whether the welfare to be maximized is real or illusory! The other is that, even if economic theory limits itself to the study of behavior, failure to distinguish between the real and the imaginary is likely to diminish the accuracy of the analysis for purposes of prediction. For while it is true that market behavior and structure at any given moment depend entirely on subjective conditions, unless the state of knowledge is unalterable what happens *through time* depends also on objective factors. Derived wants based on erroneous ideas about objective reality are less likely to persist than those based on correct ideas.

Attempts have been made to distinguish between "real" and "imagined" properties of goods by employing the test of whether the consumer's scale of preferences would be changed as a result of acquiring firsthand experience with a product, or whether he would feel "worse off" or "disappointed" after making a change.[15] This is

[13] Frank H. Knight, *Risk, Uncertainty, and Profit,* p. 157 *n.* See also pp. 183 *n.*, 185, 261–62, 339.

[14] Chamberlin, *Monopolistic Competition,* Appendix E, p. 271.

[15] See, e.g., J. E. Meade, *An Introduction to Economic Analysis and Policy,* 2d ed., p. 155.

a step in the right direction, yet has shortcomings. Unless it is specified that *all* possible alternatives are tried, the test does not tell us whether the selection is the very best possible. And even so, it could happen that a consumer, even after using a product, might fail to realize elements of superiority or inferiority in it; he might still be living in a fool's paradise, though a different one.

The concept of basic wants makes possible a more precise definition of the "objective reality" to which subjective evaluations of products must conform in order to maximize welfare. The physical properties of goods and their possible uses are part of objective reality. So are the consumer's current circumstances and his underlying tastes and values, which determine his basic wants. If a consumer had perfect knowledge and perfect powers of analysis, and sought to maximize his satisfactions, he would select the products that could satisfy these basic wants with the greatest precision and economy. Then his derived wants would be optimal—and we could say that his subjective preferences were based wholly on "real," not "fancied," considerations. This amounts practically to saying that what people "ought to want" is what they would want if they had unlimited time, resources, and expert assistance at their disposal to enable them to analyze perfectly their own tastes and all the available products.

What we have arrived at is a hypothetical indifference function —the consumer's function as it would be if all calculations and all the data on which they were based were "correct." In conceiving this function we must assume that the consumer's tastes, values, and circumstances remain unaffected by the acquisition of omniscience. In actual fact, of course, this would be most unlikely. Yet for purposes of *economic* analysis the only legitimate thing to do is to accept the consumer's aesthetic leanings and philosophy of life as they are and consider how his income could best be employed in satisfying them.

Now we have a clear-cut theoretical dividing line between "real" and "fancied." If a consumer is choosing between two brands of perfume which are identical except that Brand A comes in a plain bottle and Brand B in a fancy bottle, and if the consumer, though indifferent about the shape of the bottle, prefers Brand B because the fancy bottle has tricked him (or her) into supposing that the perfume inside must be better, his preference is based on a fancied difference. But if he prefers the fancy bottle because its shape gives him aesthetic pleasure, and is aware that the two perfumes are identical, the preference is based on a real difference.

Part Three

THE QUALITY VARIABLE AND COMPETITION

Chapter VIII

THE NATURE OF ECONOMIC COMPETITION

SO far, our inquiry into the quality elements in competitive exchange has consisted of an examination of consumer and producer behavior in situations in which no institutional restraints interfere with the freedom of producers or consumers to compete with each other. We have not yet tried to determine the "degree of competitiveness" of the markets that arise under such circumstances. Nor have we dealt with the character of the markets that arise when freedom of economic action is not complete.

It is now time that we turn to such matters. Before doing so, however, we must make certain that we understand what we mean by the term "competition." What *is* competition? Is it something that exists only in a "purely competitive" market? Is it something which is weakened or strengthened by "quality competition"? Is dissimilarity between competing products a "competitive" or an "anti-competitive" element? Is advertising an indication of competition, or of deficiencies in competition? Questions of this sort reveal the ambiguities involved in statements about competition when the term is not carefully defined.

This chapter and the next are devoted to a clarification of this crucially important concept.

1. THE NEED TO DEFINE COMPETITION

A definition of competition is surely necessary for any analysis which seeks to isolate the competitive forces in capitalist markets. Yet curiously, most modern economists write about competition without defining it, presumably because they believe its meaning self-evident. As Frank Knight has said, "The critical reader of general economic literature must be struck by the absence of any attempt accurately to define that competition which is the principal subject under discussion." [1] Even when an attempt is made at a definition, what is defined is not competition itself, but a particular kind of competition, such as "pure" or "perfect" or "simple" competition.

Moreover, such definitions are usually of a sort that does not tell us what competition *is,* but simply describes the conditions under which it exists, or its consequences. Perfect competition, we are told, is "a state of affairs in which the demand for the output of an individual seller is perfectly elastic." [2] Simple competition is defined as the kind that exists where "each seller produces as much as he can at the ruling market price, and does not restrict his output in the hope of causing that price to rise." [3] To persons who did not already have a comprehension of the competitive process, definitions of this sort would be quite unsatisfactory.

2. ECONOMISTS' CONCEPTS OF COMPETITION

Contemporary economists are inclined to view competition as that which is found in its purest form and highest degree in "pure" or "perfect" competition. Their concept embraces two principal ideas as to market characteristics considered to be indications of competitiveness. In the first place, competition in selling is said to mean the offering of identical goods by different suppliers; and in buying, the bidding for identical goods by different buyers. If the goods are not identical the situation is not "competitive" or is only "imperfectly" competitive. The second chief competitive element, according to accepted theory, consists in there being a sufficiently large number of rivals so that no one of them can exert an appreciable influence on the price. On these two cornerstones—*identical* products and *many* participants—has been built the concept of

1 Frank H. Knight, *Ethics of Competition,* p. 49.

2 Joan Robinson, "What Is Perfect Competition?", *Quarterly Journal of Economics* XLIX (Nov., 1934), 104.

3 A. C. Pigou, *The Economics of Welfare,* 4th ed., p. 213.

"pure" competition. For competition to be "perfect" there must also be, according to most formulations, perfect and costless mobility, divisibility, and knowledge. Then the acme of competition is reached.

These ideas, though firmly accepted by many today, are not deeply rooted in the soil of economic thought. The classical economists did not conceive of the competitiveness of a market as depending upon the number of sellers or the homogeneity of the product, but upon the amount of freedom enjoyed or restraint imposed. Theirs was the concept of *free competition.* Adam Smith described a competitive society as being one in which there was "perfect liberty" or "where things were left to follow their natural course." [4] Ricardo and Jevons also looked upon competition in this light.[5]

It was Augustin Cournot who started economists thinking in terms of monopoly, duopoly, and increasing numbers of sellers. His conclusion that profit is a function of the number of sellers, being wholly eliminated only in "unlimited competition," gave subsequent students of value theory a precise mathematical yardstick by which the degree of competition might be measured.[6] Yet as recent an economist as Marshall viewed competition as something to be contrasted with "the yoke of custom and rigid ordinance" and to be described in such terms as "freedom of industry and enterprise" or "economic freedom" or "the undisturbed action of free competition." [7] Not until the twentieth century did pure or perfect competition completely displace free competition in the theoretical literature as the criterion of a competitive market.

Do the notions of freedom and perfection represent two alternative, conflicting concepts of the underlying nature and meaning of competition? Strictly speaking, no. What economists really mean by perfect competition, even if they do not clearly say so, is the set of conditions under which competition leads to the "ideal" state of

[4] Adam Smith, *The Wealth of Nations,* Modern Library ed., Bk. I, ch. vii, pp. 56, 62; ch. x, p. 99.

[5] David Ricardo, *Principles of Political Economy and Taxation,* Gonner ed., p. 7; W. Stanley Jevons, *The Theory of Political Economy,* 4th ed., p. 86.

[6] Augustin Cournot, *Researches into the Mathematical Principles of the Theory of Wealth, passim,* esp. p. 90.

[7] Alfred Marshall, *Principles of Economics,* pp. 12, 10, 35.

affairs in which goods are produced at minimum unit cost and priced at cost. The enumeration of these conditions does not tell us what competition is, but only what must accompany it to make it accomplish certain results. And when these conditions are met it is not *competition* that is maximized but the *beneficial results* of competition. Yet the fact remains that full freedom of competition does not invariably tend to bring about the conditions associated with pure or perfect competition. In fact the opposite is usually true; due to nature's abhorrence of sameness and man's capacity for thinking up ingenious variations of products, pure competition is extremely unlikely to occur unless some standardization or regulation of quality is enforced on all who enter the market.

A number of modern economists have sought to describe competition in broader, looser terms than those of formal theory. In most cases their immediate and primary objective, like that of the theoretician, has been simply to state the kind of competition or the accompanying conditions needed to insure desirable results. This seems to be true of most formulations of "workable competition" and "effective competition" as well as of discussions of the kind of competition that antitrust laws are, or ought to be, aimed at. Nevertheless this segment of the literature is a valuable source of ideas about competition itself.

One significant idea advanced by J. M. Clark, Corwin D. Edwards, and others is that competition means the availability of alternatives. The word "availability" here implies two things; first, that the alternatives exist; and second, that the participants have the power to choose freely among them. Clark, for instance, defines "competition in price between business units" as "rivalry in selling goods . . . under conditions such that the price or prices each seller can charge are effectively limited by the free option of the buyer to buy from a rival seller or sellers of what we think of as 'the same' product. . . . This . . . definition . . . focuses attention on a crucial point which is sometimes neglected—namely the nature of the option actually open to the buyer." [8] Edwards characterizes the competition "that prevails in actual competitive markets" as

[8] John Maurice Clark, "Toward a Concept of Workable Competition," *American Economic Review*, XXX (June, 1940), 243. See also his *Alternative to Serfdom*, p. 70.

consisting of "access by buyers and sellers to a substantial number of alternatives" and "their ability to reject those which are relatively unsatisfactory." [9] Clare E. Griffin's discussion of the distinction between competition and the bargaining relationship [10] reveals a similar concept of competition. To him, there is "little competition" if "there are few, if any, practical alternatives to the buyer and to the seller." This idea therefore goes beyond the concept of competition as being freedom of entrepreneurial action. Freedom to create new alternatives is enough to assure *potential* competition. But for *actual* competition to exist the alternatives must actually be created. On the other hand, it is a less restrictive concept than that of pure competition. "Suppliers and customers do not need to be so numerous that each trader is entirely without individual influence," states Edwards, "but their number must be great enough that persons on the other side of the market may readily turn away from any particular trader and may find a variety of other alternatives." [11]

A second and closely related idea is that the alternatives need not be, in fact preferably should not be, identical in quality. According to Edwards, "Competition implies . . . that there are alternatives available in the market in business policies, whether toward prices, production, or the kinds of goods and services which are furnished. . . . In markets where all producers are united . . . in a common policy toward the terms of sale or toward the characteristics of their products, effective option by the buyer is . . . destroyed." [12] M. A. Adelman also views competition among buyers and sellers of a homogeneous product as limited and incomplete. "The consumer is not benefited by a choice between Tweedledum and Tweedledee; he needs a wider market, which includes at least one real alternative. . . . A sufficient number of alternatives open to any given buyer or seller are necessary, including alternatives

[9] Corwin D. Edwards, *Maintaining Competition*, p. 9.

[10] Clare E. Griffin, *Enterprise in a Free Society*, p. 279. This quotation and others from Griffin's *Enterprise in a Free Society* are reprinted with the permission of Richard D. Irwin, Inc.

[11] Edwards, *Maintaining Competition*, p. 9.

[12] Corwin D. Edwards, "Can the Antitrust Laws Preserve Competition?", *American Economic Review: Papers and Proceedings*, XXX (March, 1940), 170–71. See also *Maintaining Competition*, p. 9 *n*.

in the type of goods ('stripped' versus begadgeted models, for example)." [13] Griffin describes "competition in practice," as distinguished from "the perfect competition of economic theory," as characterized by "the wide variety of qualities and types of goods which are vying for the consumers' favor," and notes that this "appears to the businessman to be the highest development of competition." [14] Schumpeter has argued that "in capitalist reality as distinguished from its textbook picture" competition in price among sellers of the same commodity, produced by the same methods, is far less effective and less powerful than the competition that arises when qualitatively different alternatives are pitted against each other. "In the case of retail trade the competition that matters arises not from additional shops of the same type, but from the department store, the chain store, the mail-order house and the supermarket." [15]

A third idea is that competition is something more than the mere existence of a situation in which rival, dissimilar alternatives are available. It is a type of action induced by such a situation, the action being of a nature that alters the relationship between the rival participants. In other words, competition is a dynamic process. Its central element, according to Clark, is "offering the other party a bargain good enough to induce him to deal with you in the face of his free option of dealing with others." [16] If the bargain is not already good enough it must be made so. Sellers are put under pressure to improve their offers by buyers who shift their purchases away from those firms whose offers seem inferior.[17] The result is movement, change, the alternative creation and destruction of differentials between rival offers. As Griffin describes it, "competition of producers" consists of the "effort of each producer to get or to keep patronage which might go to another. These efforts take the form of striving to make the offer more attractive to the buyers than the offers of competitors. This improved attractiveness of the

[13] M. A. Adelman, "Effective Competition and the Antitrust Laws," *Harvard Law Review*, LXI (Sept., 1948), 1295, 1303.

[14] Griffin, *Enterprise in a Free Society*, p. 283.

[15] Joseph A. Schumpeter, *Capitalism, Socialism, and Democracy*, pp. 84, 85.

[16] John Maurice Clark, *Alternative to Serfdom*, p. 70.

[17] Corwin D. Edwards, "Can the Antitrust Laws Preserve Competition?", *American Economic Review: Papers and Proceedings*, XXX (March, 1940), 170.

offer may be a lower price; or it may be improved quality, a more attractive design, a more useful or attractive package, greater convenience of location of the point of sale, and assurance of dependability that comes from a long-established record, improved after-sales service—such as repairs or adjustments, and many other features." [18] Adelman likewise describes competition in terms of change rather than of position: "The pursuit of business advantage in a competitive market takes the form of reductions in price, improvements in quality, and a constant search for cost reductions and innovations." [19]

For competition in this sense to occur, the situation must be either one of actual disequilibrium or one in which existing relationships between buyers and sellers can be upset at any moment. It must also be such that producers have some scope for decisions regarding price, quality of product, and terms of sale. "The terms have to be worth your while—that is a bottom limit—and the top limit is set by the other party's freedom to go elsewhere if anyone else offers terms more attractive or more advantageous to him than yours are." [20] Whenever these two limits coincide, competitive action in the dynamic sense is no longer possible.

Hayek views competition in this light. According to him,[21] "competition is by its nature a dynamic process whose essential characteristics are assumed away by the assumptions underlying static analysis." It is "a process of the formation of opinion" and "the process by which the state of affairs is brought about which [economists] merely assume to exist." It withers away when perfect knowledge and equilibrium are attained; " 'perfect' competition means indeed the absence of all competitive activities."

Adam Smith too conceived of competition as a process that arises in free markets when disequilibrium occurs. "When the quantity of any commodity which is brought to market falls short of the effectual demand . . . a competition will immediately begin

[18] Griffin, *Enterprise in a Free Society*, p. 285.

[19] M. A. Adelman, "Effective Competition and the Antitrust Laws," *Harvard Law Review*, LXI (Sept., 1948), 1303.

[20] John Maurice Clark, *Alternative to Serfdom*, p. 70.

[21] Friedrich A. Hayek, *Individualism and Economic Order*, pp. 94, 106, 93, 96, 103.

among them, and the market price will rise." [22] What Hayek and Smith are saying is that competition becomes *active* only in disequilibrium, and remains *dormant* in "competitive" equilibrium. If competition is viewed as a dynamic process, this conclusion is inescapable.

3. NONECONOMIC COMPETITION

Competition is of course something that exists out of the economic world as well as in it. Students compete for prizes, scholarships, and grades. Athletes compete individually and in teams. Competitive civil service examinations determine (in theory, at least) the person best fitted for each government job. Courtship operates on a competitive basis. In the arts, committees of judges are forever selecting some opera, or budding violinist, or painting, or architectural design out of a group of competitive entries. In democratic politics competition is regular and frequent; for an election is nothing more nor less than a competition in which the candidates are contestants and the voters function as a "committee of judges." People frequently speak of "the competition of ideas"; competition of this sort exists in almost every realm of thought and activity. It may help our understanding of economic competition if we examine these other forms of competition in an attempt to find characteristics possessed in common.

Perhaps our best starting point is a dictionary or encyclopedia definition. *Webster's New International Dictionary*, Second Edition, Unabridged, defines *compete* as "to contend emulously . . . to stand comparison, as with respect to fitness or value," and defines *competition* as "a contest between rivals." *The Century Dictionary* gives as the meaning of *competition* "the act of seeking or endeavoring to gain what another is endeavoring to gain at the same time . . . strife for superiority; rivalry." In *The Encyclopedia of the Social Sciences* Walton H. Hamilton describes competition as "the key word in an account . . . of how rivalry . . . comes to promote organization. It is by competition—whether of persons, firms, industries, nations, races, beliefs, habits or cultures

[22] Smith, *Wealth of Nations*, Bk. I, ch. vii, p. 56.

—that the fittest survive. . . . Competition is . . . a process of selection."

Here we find several significant ideas: first, that competition involves a contest, that is to say a trial or testing out of rival plans, ideas, or skills, in order to determine something that is not determinable in advance; second, that it includes the process of judging—that is, of comparison, evaluation, and selection—which may be done either by persons or through some impersonal mechanism; third, that selection is made on the basis of "fitness" or "value"—that is, suitability or high ranking in terms of the conditions of the contest—which implies some predetermined method of evaluation or "rules of the game"; fourth (and this is implied by the fact that the counterpart of selection is rejection), that competition is a weeding-out process that serves to eliminate the "unfit"; and fifth, that it involves the striving of the contestants to outdo each other in order to reach a position in which the judgment will be favorable.

These ideas are found in other analyses of competition. Cooley, for example, has described personal competition as a "selective process" which "decides the function of the individual" on the basis of his "personal qualifications" through a "process of comparison," the decisions being made by "selective agents" of some sort.[23] John Dewey also conceives of competition in terms of alternatives, trial, comparison, and judgment, and views competition as something that is involved in all reflective thinking.[24] Kingsley Davis emphasizes the fact that competition is a struggle to excel, not to extinguish, one's rivals. "In contrast to conflict, which aims to destroy or banish the opponent, competition simply aims to out-do the competitor in achieving some mutually desired goal. It is thus a modified form of struggle. It implies that there are rules of the game to which the competitors must conform and that behind these rules, justifying and maintaining them, is a common set of values superior to the competitive interest. It also implies an absence of coercion. The rules are so arranged that the ends

[23] Charles H. Cooley, *Sociological Theory and Social Research*, pp. 164, 165, 168.

[24] John Dewey, *Human Nature and Conduct*, p. 190; *How We Think*, p. 103.

must be obtained by other methods than fraud or physical force." [25]

This distinction between conflict and competition is blurred in some competitive sports, such as boxing and wrestling, which are tests of excellence in physical combat; yet on the whole it seems an important and useful one. The competitive world has occasionally been likened to the jungle; in both, it has been pointed out, a fierce tooth-and-claw struggle for survival takes place. There is some truth in such an assertion. Competition does occur in nature's evolutionary processes. Darwin's phrase "natural selection" and Spencer's "survival of the fittest" both refer to nature's omnipresent weeding-out process, which eliminates the unfit, leaving the fittest to flourish and multiply. It is also true that competition is often cruel and ruthless, as when losing involves not only elimination from the contest but personal hardship or humiliation or even extinction. But there is a basic difference between the direct efforts of two jungle inhabitants to destroy each other and the efforts of two jungle inhabitants to gain simultaneously a coveted article of food, even though the winner in the second case may indirectly cause the extinction of the loser through starvation. When a Congressman abandons oratory and resorts to his fists, or when attempts are made to drug a race horse, competition gives way to conflict.

Outside the economic sphere, then, the word "competition" means a type of activity that involves contestants who are pitted against each other, some common goal sought by them, efforts on the part of each to achieve superiority in attaining the goal, some method of judging superiority, judges or judging mechanisms to do the judging, the selection of one—or perhaps several—among the contestants, and a rejection of the others.

4. A SYNTHESIS AND A DEFINITION

Let us now try to put together these various ideas and arrive at a definition of economic competition which does not do violence to the generally accepted meaning of the word, yet is applicable to market situations and suitable for purposes of economic analysis.

25 Kingsley Davis, *Human Society,* p. 162.

It is clear that economic competition, at least in some of its aspects, is not fundamentally different from noneconomic competition. It involves contestants and judges. In most markets, the sellers are the contestants—they take the initiative in offering bargains—while the buyers act as judges, selecting bargains they consider superior and rejecting those believed to be inferior. Each purchase represents the outcome of a contest, the selection of a winner. In factor markets the roles are usually reversed: the firms which are seeking to buy the services of labor or capital quote terms on which they are willing to exchange, and thus play the role of competitors, while the individuals or firms able to provide these services evaluate the bids.

Economic competition also consists of striving toward a common goal; the competitors are doing "the same thing," although perhaps in different ways and with different degrees of success. Orthodox theory, it is true, finds this to be the case only in markets that are purely competitive. But if we adopt the viewpoint that dissimilar products can serve the same purpose—that is, are qualitatively different means to the same end (the attainment, let us say, of some definable experience or activity or other goal sought by buyers) —sellers of differentiated goods can clearly be regarded as participants in the same contest.

Economic competition may also be viewed as a weeding-out process. Among goods, as among creatures of the jungle, or players in a tennis tournament, a continual struggle goes on for the survival of the fittest. The history of the automobile industry is littered with the names of now-defunct cars—makes which have been rejected by the buying public in favor of cars that for one reason or another they liked better. In many competitive industries high-cost producers have similarly been weeded out. Throughout the business world a continual process of competition for survival or preferential place goes on—between persons (for example, rival salesmen), between brands (Luckies versus Chesterfields), between kinds of goods (orange versus grapefruit juice), between types of services (highway versus railway transportation), between ways of spending money (a vacation trip versus a new car). Wherever there is opportunity for substitution there is competition.

On the other hand, economic competition differs somewhat from noneconomic competition.

For one thing, in many market transactions each participant is both a contestant and a judge. The purchaser of a house, for example, typically acts as a judge in appraising the merits of a number of competing alternatives, then competes with other purchasers in attempting to obtain the house of his selection on the best possible terms, while the seller of the house judges the attractiveness of the various offers and also competes with other sellers in attempting to get buyers to pass favorable judgment. In fact whenever there are several sources of supply and several possible outlets for each supplier and each exchange is considered a unique individual to be bargained for separately (as in the organized commodity and securities markets and in markets for real estate, used cars, used yachts, construction contracts, and major repair jobs), economic competition has this dual character.

Is it the plurality of both buyers and sellers and the reciprocal relationship between them that accounts for the duality of economic competition? Partly, it would seem. But it seems to be due also to the fact that the price element, lacking in noneconomic competition, ordinarily forms a necessary part of economic competition. Whenever price does not enter into economic competition, as in the rivalry of radio and television programs, which are offered free to owners of receiving sets, or in competition for civil service jobs, where the price is fixed in advance, the duality disappears.

This leads us to consideration of another, even more striking difference. In markets where price is the sole consideration—the so-called purely competitive markets—economic activity seems to bear almost no resemblance to the activity of contestants and judges in games and prize contests. In the textbook explanations, certainly, what occurs is not a weeding-out process (since all sellers are assumed to be offering identical products and to have identical costs), but rather a balancing or adjusting process, a ceaseless experiment to find equilibrium. Traders who seek to buy units of a standardized commodity do not judge their quality, or reject the unfit units, or give preferential place to any individual seller's supply. That type of activity would be fruitless, since the market organization has in some way already assured the homogeneity of the product.

Nevertheless in pure competition a process of selection and rejection does take place. This becomes evident if we turn our attention away from equilibrium positions. And interestingly enough its net result, like that of a tennis tournament, is a weeding-out of the unfit. The contestants, however, are not the products offered by the competing sellers; they are the bid and asked prices.

This fact is perhaps obscured by our tendency to regard competition as something that always takes place between *people*. True, most forms of noneconomic competition are contests to determine superiority in personal ability, and economic competition, too, often involves tests of entrepreneurial skill. But competition need not be between human beings. In horse racing the actual contestants are the horses, not their owners. So it is with the competition between imperfect substitutes: it is the products themselves, not their sellers, which are being pitted against each other.

Let us now ask ourselves, who or what are the competitors in a purely competitive market? The units of the standardized commodity, about which the buyers are completely indifferent, since comparison is unnecessary? Or the traders, who are considered equally acceptable to deal with? Or the prices? Clearly, the last. In the organized commodity and securities markets of the actual world buyers are continually rejecting prices which they believe to be too high, and selecting the lower of any pair of contesting prices. Sellers, quite naturally, are continually doing precisely the reverse: for them the preferred price is always the highest. Thus two simultaneous processes of competitive selection are constantly being balanced off against each other. But the objects of comparison and selection are not rival products but rival prices.

The price element, then, introduces a new dimension into competition. When competitors are offering their creations or services without charge (as in the competition of ideas in a debate) or for some fixed, predetermined reward (as in a prize contest), the only relevant variable is the quality or worth of each contestant's offering. It is the best idea, not the cheapest idea, that wins. But when price tags are attached to ideas or services or products, it is the best *bargain* that wins. How good a bargain anything is depends upon both quality and price; these two elements *compounded together* form the basis for evaluation of winning contestants in the market place. Only when differences in quality have been eliminated by standardization does "cheapest" necessarily coincide with "best."

It thus appears that economic competition, like noneconomic competition, can be described as a process by which rival alternatives are proffered, tested, evaluated, and selected or rejected. Economic competition is usually more complex than noneconomic competition, partly because of the reciprocal relationship between buyers and sellers and partly because the evaluation process which takes place in most markets consists of both a weighing of product against product and a weighing of price against price.

Let us now define economic competition: It is a contest—or, more usually, a succession of contests—in which independent sellers enter products of their own choosing, at prices of their own choosing, for appraisal and purchase by independent buyers, the products being substitutes for each other in the sense of being alternative means to the attainment of some activity or experience, the buyers being free to select or reject any bargain offered and to make their own offers of terms, and all participants being guided in their decisions by their own conceptions of their best interest.

The rival products may be either perfect or imperfect substitutes for each other; and so may the rival prices. The independence of the sellers and buyers implies (in a world lacking uniformity and constancy of tastes, perfect knowledge, and instantaneous adjustment) some dissimilarity some of the time between rival products and between rival prices. Freedom to select or reject any bargain and to alter bargains may be limited by legislation, custom, or closed-mindedness; the essential condition is that some significant amount of freedom remain and be exercised. This freedom includes freedom to alter the qualities of products as well as the terms on which they are offered. The voluntary character of the participation and of the decisions made by the participants implies a desire on the part of the competitors to be judged favorably, and to alter products, offers, and bids whenever it seems advantageous to do so.

Competitive *striving*, though not a part of the definition just given, may certainly be inferred from it.

This definition classifies sellers as contestants and buyers as judges. In some markets (for example, most factor markets) this classification may seem out of place. In order to make our definition general, therefore, we must either regard the roles as capable of being reversed or regard the judges as occasionally taking on an additional function: assuming the initiative in suggesting to the contestants how they may revise their conduct so as to obtain a more favorable judgment.

This concept of economic competition is broader than the usual theoretical concept. It does not stipulate the capitalist institution of private enterprise for pecuniary profit, and thus embraces the activities of nonprofit organizations (for example, the college education industry). It does not stipulate large numbers, and therefore embraces duopoly-duopsony situations. (Yet it does stipulate independence, and so rules out collusive oligopoly.) It does not stipulate a homogeneous product—in fact it suggests that there is something incomplete about competition in which quality differences are lacking. The crucial test is the availability of alternatives. If the alternatives are limited, so that feasible alternatives which might prove superior to those already entered are denied admission to the contest, the resulting competition is limited; but only when there are no alternatives at all, or when contestants are not free to vary their conduct as a means of trying to better their previous performances, is competition lacking.

This concept, aside from the label chosen to identify it, seems useful as an identification of a type of market situation which is often but not always found in the real world, and which has definite welfare implications. And the label "competition" seems appropriate, partly because it is in accord with everyday usage of the word and thus reduces the likelihood of misunderstanding, and partly because its relationship to the concept of pure competition turns out to be an interesting one for purposes of economic analysis.

Chapter IX

PRICE AND NONPRICE ASPECTS OF COMPETITION

ARMED with a working definition of competition, we are now better able to reexamine two well-known concepts in modern economics which occupied our attention in Chapter I: "nonprice competition" and "pure competition." How do they fit into the picture?

1. IS NONPRICE COMPETITION COMPETITIVE?

Current economic theory generally treats nonprice competition—that is, the effort to attract customers through quality differences, quality changes, and advertising—as anti-competitive elements indicative of departure from a wholly competitive situation. What light is thrown on this question by the analysis just made?

It has already been shown that comparison of the various competing alternatives is an essential part of the competitive process. In noneconomic competition this always takes the form of quality comparison, for the winner is always selected on the basis of qualitative superiority. In economic competition a new element enters: price comparison. But quality comparison is not thereby eliminated. No product can be considered satisfactory or acceptable in quality unless it stands the test of comparison with other quality alternatives. Even in the "purely competitive" markets someone, somehow, must undertake the job of quality comparison in order to determine that the competing products are in fact identical.

Moreover, in the actual world there is not a single multifirm industry in which differences do not occur and in which quality comparison is not a live issue. It would certainly be straining the meaning of the word "competition" to define it in such a way as to include the competition between dissimilar athletes, architectural designs, and political candidates, yet exclude the competition that arises because of differences in quality among products. All of these are examples of quality competition.

Without quality differences, quality competition would serve no useful purpose and would be nonexistent. We can therefore say that quality differences are a prerequisite for this kind of competition, and so introduce a competitive element that would otherwise be lacking. Far from diminishing the degree of competition in a market, they enhance it.

Quality changes too are competitive. When the rival performances of contestants are not subject to change, there is no reason, aside from imperfections in the ability of the judges, for holding new contests and passing fresh judgment. If the qualities of products and the bid and asked prices in any given market remain constant, yesterday's competition is sufficient to establish once for all time the relative positions of the contestants in the minds of the judges. But if the performances of the contestants are variable, recurrent contests are needed. The continuance of quality competition therefore hinges largely upon quality changes (that is, product variation, new models, new products).

We are now better able to perceive why contemporary theories tend to regard quality differences as monopolistic. In static equilibrium analysis the qualities are by hypothesis immutable, and each seller's product in a monopolistically competitive market is therefore inimitable, so competitive striving for quality superiority would be fruitless. Under such conditions quality differences are indeed monopolistic. Even when expansion of demand brings new firms and new varieties, encroachment on the positions of the monopolists already in the field is regarded as impossible; the newcomers, like homesteaders, are forced to seek unpopulated areas. But this is mere abstraction. Quality immutability or protection against encroachments exists only occasionally in the real world, as when a superior product is kept unapproachable through patents or other institutional barriers or through natural obstacles. Typically, qualitative superiority is a challenge to competitors to emulate or outdo the successful producer's performance. Then it is better described as a competitive element. It seems correct to say that whether quality differences and changes are competitive or monopolistic depends first on whether they are open to imitation, and, in those cases where imitation is barred, whether they spur competitors to erase the superiority by counter-innovation.

Is advertising competitive or monopolistic?

A considerable body of opinion holds that it is monopolistic. Yet advertising that is informative facilitates quality comparison, and thus sharpens quality competition, by improving buyers' awareness of alternative ways of satisfying existing wants or of obtaining new, untried experiences. It makes buyers more fully informed about market situations, reminding them of the existence of alternatives which they may have forgotten about and pointing out quality differences that may have escaped their attention. It also enhances price competition whenever it informs buyers of price developments which alter the relative attractiveness of the various alternatives. By doing these things informative advertising helps both quality competition and price competition to perform their functions more efficiently.[1]

Viewed in isolation, the informative advertising activity of each business firm apparently increases monopolistic control by building up consumer preference for that particular variety, thus making the firm's demand curve less elastic. But the net effect of industry-wide informative advertising is likely to be the opposite. In the absence of advertising, investigation is likely to be more difficult, more costly, and less fruitful, and therefore less indulged in, so that buyers are less likely to know of suitable alternatives. Their attachments will consequently be fairly strong. If informative advertising is introduced into such a market it will break down preferences based on inadequate knowledge. And if all firms advertise, any individual advertisement is as likely to weaken as to strengthen buyers' attachments. For while A's advertising tends to

[1] See Chamberlin, *Monopolistic Competition*, pp. 118, 119.

strengthen the attachments of those who currently prefer A's product, the advertisements of B, C, D, etc., tend to weaken them.[2] On the whole we may conclude that industry-wide advertising is more likely to weaken them, since in a world of imperfect knowledge the acquisition of knowledge of close substitutes previously unknown has the same economic effect as the production, in a world of perfect knowledge, of close substitutes previously unproduced: it tends to make demand curves more elastic.

Noninformative or misinformative advertising, on the other hand, tends to diminish the effectiveness if not the degree of competition by distracting buyers from their task as judges or by deceiving them as to the merits of the contestants. The resulting competition may be just as keen, but its basic purpose—that of determining the "best" alternative—is thwarted. If intelligent evaluation is considered to be an essential part of the competitive process, competition itself has been weakened. An exception may be noted: some noninformative advertising neither distracts nor misinforms, but is simply irrelevant; such advertising leaves the state of competition unaffected.

In short, nonprice competition enhances the competitiveness of markets unless it is accompanied by barriers to imitation or by deceptive or diversionary advertising techniques.

2. HOW COMPETITIVE IS "PURE COMPETITION"?

We have found quality competition to be a genuine form of competition, distinct from price competition, and capable of occurring either along with price competition or without it. When occurring alone (as under conditions of perfect price rigidity), it reveals one form of competition in isolation. Thus quality competition, though competitive, is only incompletely so.

Price competition is likewise a form of competition which can occur either along with quality competition or without it. It too is an example of partial competition, for if quality alternatives are lacking a genuinely competitive element is omitted. "Pure

[2] *Ibid.*, pp. 166–67, 187.

competition" is the name given to such a competitive situation, in which price competition exists in isolation, quality differences being eliminated by the standardization of the product. If our analysis has been correct, such a situation is not a fully competitive one, as is usually implied in textbook chapters on pure competition; the competition, though "pure," is incomplete.

This does not mean that theorists are in error in selecting the conditions of pure competition as the basis for a theory of competitive price. In studying price determination it is necessary to isolate the price variable and to hold "other things" (including the quality variable) constant. Such a procedure makes possible a precision in solving problems of competitive price otherwise unobtainable. Pure competition is an analytical device, an abstraction, a mental picture obtained by looking at competition solely from the standpoint of price, which shuts out from the economist's vision what cannot possibly be omitted in any faithful report on the competitiveness of a market: the quality competition that occurs whenever there is choice between imperfect substitutes. A commodity traded in a "purely" competitive market may or may not face competition from other commodities capable of serving the same purpose. Price theory does not concern itself with this question, for whether it does or does not is immaterial for purposes of price determination.

But what is essential to a study of price is not necessarily appropriate for the theory of economic welfare. If other things beside price are relevant in the determination of welfare, price theory is an incomplete foundation for welfare theory. Nor can it be assumed as a matter of course that the conditions which isolate the price variable are the conditions of fullest competition.

Just as quality competition can be described as a contest between competing qualities to decide preferability, so price competition can be viewed as a contest between competing prices. On the sellers' side this means the effort to undersell one's rivals; on the buyers' side, to outbid one's rivals. Both quality and price competition imply comparison, and comparison implies differences. (When two things are known to be alike, comparison is unnecessary, and the only rational attitude is one of indifference.) Price differentials thus seem to be a prerequisite of price competition. Indeed the price competition of the real world, as evidenced in auction sales, "chiseling" tactics, and price "wars," typically consists of the creation of price differences followed by their removal or the creation of new differences.

If this is so, how competitive is a "purely competitive" market? In such a market, we are told, there can be only one price. (As the argument is usually stated, nobody would buy from a seller who demanded a higher price than other sellers, and nobody would be willing to sell for less, being able to dispose of unlimited amounts at the ruling price. Therefore nobody does!) This means that all competitors are offering the same commodity at the same price. The result is analogous to a race in which all contestants cross the finish line at the same instant, resulting in a "dead heat." There are no winners, and no losers. The judges view the performances of the competitors with complete indifference. Selection must perforce be at random. What kind of competition is this?

It is important to realize that the description just given is one of a purely competitive market *in equilibrium*. This is sometimes not made explicit in price theory, it being taken for granted that the reader is aware that the theory is concerned, not with the process of competitive price adjustment, but with the end result. The process is glossed over by assuming instantaneous adjustments to instantaneously known changes, so that a progressive society can be pictured as moving effortlessly from one equilibrium position to another.[3] But in a world such as ours, in which there is continual change, imperfect knowledge, disagreement, and lagged adjustment, equilibrium positions are never reached but only approached. The ever-repeated "dead heats" of the theoretical model could actually occur only in that inconceivable society, the stationary state. And if they did occur, the result could hardly be called competition in the sense in which the term is generally understood. When every entry is as good as every other, there are no inferior

[3] See Frank H. Knight, *Risk, Uncertainty, and Profit*, p. 82: "We need not dwell upon the process by which fixed rates of exchange among all commodities will be established. . . . If intercommunication is actually perfect, exchanges can take place at only one price."

entries to be eliminated, and there is no need for discrimination or selective choosing on the part of the judges. Competition becomes inactive.

Price competition, then, is not the ceaseless round of exchanges on identical terms that occurs under equilibrium conditions but is instead an institutional device for bringing about price movements toward equilibrium. Without competition, a uniform price above equilibrium might persist; buyers would be exploited, and resources would not be allocated in the most effective way. Price competition is a mechanism to correct this condition by bringing into existence alternative prices from which to choose. Price competition is inaugurated by the creation, by someone, of a price differential. In highly organized markets this differential will be between two simultaneous offers or bids; in more loosely organized markets it may be between the prices at which the commodity is actually traded. The differential puts rivals at a disadvantage, and forces them to "meet" the competition by altering their own prices. Note that the phrase "meeting competition" implies that the creation of the advantageous differential was a competitive act. Note also that price is not accepted as a datum by all buyers and sellers —otherwise no change could occur. There is of course a time sequence to these price adjustments; the pattern is one of challenge and lagged response. If equilibrium is reached no further differential can be advantageous to anyone, and the process stops.[4]

This view of price competition as a dynamic process leads to the conclusion that competition tends to die down in neighborhoods of equilibrium and to spring up or become more active when a disequilibrating change occurs.

How competitive is "perfect competition"? Actually, perfect competition is a paradox; it can hardly be called competitive at all. Its very conditions obviate the need for competition. Competition is a test, a trial, the outcome of which provides the basis for evaluation. If the outcome were known in advance, the contest would be unnecessary. In a perfectly competitive market everyone is assumed to know at every instant all he needs to know. There

[4] See John Maurice Clark, *Preface to Social Economics*, p. 111; *Studies in the Economics of Overhead Costs*, pp. 417–18.

are no lags in the intercommunication of knowledge. There can be no disagreement. In actual markets whatever approximation to perfect knowledge is achieved comes as the *result* of the ceaseless round of contests which enable judgments to be revised and sharpened. But in those hypothetical markets where "competition is perfect" the traders are supposed to possess this knowledge already. It would be a foolish waste of time to resort to a process whose purpose is to discover something already known. Rational traders intent on maximizing profits would never do such a thing! We must conclude that competition—like mystery fiction—could have no place in an omniscient society.

Again we come face to face with the idea that competition is not so much a state of affairs as a method of altering the existing state of affairs. It is, in particular, the method by which an economy in disequilibrium finds its way toward equilibrium. Any society that has already reached that happy state has no need for active competition. It is because equilibrium is so elusive in our flesh-and-blood world that competition plays so dominant a role in capitalist societies.

3. THE TWO-DIMENSIONAL CHARACTER OF ECONOMIC COMPETITION

Price competition and quality competition have been shown to be two species of competition, either or both of which may be found in any given market. But they are more than that. They represent the two dimensions of competitive evaluation. To measure the worth of any proposed bargain, it is necessary to measure both what one receives and what one gives up. One measurement relates to the merits of the product (that is, its quality), the other to the size of the payment (that is, the price). It would be absurd to neglect either. In some competitive situations both dimensions are variable; in "monopolistic competition," for example, it is customary to speak of the price variable and the quality variable. In other situations one of the dimensions is held constant and uniform, as is quality in many wholesale commodity markets, or price in the competition between most big city newspapers. But in every case both dimensions exist. Whenever the quality dimension is identical for all available alternatives, no choice of quality is possible, and quality competition is lacking or dormant. Likewise, whenever the price dimension is identical for all available alternatives, no choice of price is possible, and price competition is lacking or dormant. (This part of our argument rests on the idea that "opportunity" to choose between identities is not really an opportunity at all. "You can have either six or half a dozen of these" is certainly an example of pseudo-choice!) But though choice is not always possible, comparison is always necessary—if for no other reason, simply to determine whether or not opportunity for choice exists. Thus both the price and the quality dimensions, whether constant or variable, whether competitive or standardized, are indispensable elements in the evaluation of comparable bargains.

Thus we see that market competition is a "two-dimensional" phenomenon—a compound of two independent elements, quality comparison and price comparison. Either element may be rendered inactive by being held constant, but it cannot be done away with.

A digression to the faraway field of music may help to clarify this point. The relationship between price and quality is analogous to the relationship between melody and rhythm. Any tune (that is, any succession of musical tones actually played or sung) is compounded of two elements: melody and rhythm. The first element consists of the changes or repetitions in pitch (or, as the physicist would say, in the frequencies of the sound waves). The second element consists of the durations in time of the various tones which are played in succession. The melody element is measured in vibrations per second, the rhythm element in time units.

Each element may be either variable or constant. Sometimes a portion of a tune may consist of a constantly repeated monotone; or the tones, though varied in pitch, may be of equal duration. But it would be inaccurate, in the first instance, to say that the tune had *no* melody; actually what it has is a *monotonous* melody. Likewise, when every tone is of equal duration or when the time intervals conform to no discernible pattern the rhythm element is not *absent;* it is merely monotonous or formless.

In analyzing a piece of music it is often useful to isolate one of these elements, *as if* the other were nonexistent, in order to examine its characteristics. But holding rhythm constant does not make a tune more "musical" or even more "melodious." Since a tune's aesthetic character is compounded of the aesthetic qualities of both its elements, it would be incorrect to judge it on the basis of the melody element alone, or to give all kinds of rhythm a negative value, so that one tune is considered less "pure" than another if it contains a greater "alloy of rhythm."

In much the same way the economist must distinguish between the *pure competition* or *imperfect competition* of price theory as a tool of analysis, an abstraction, a one-dimensional yardstick by which characteristics of a market are evaluated solely with reference to the price element in competition, and *a competitive situation,* which is something concrete that cannot be fully described without taking cognizance of both its dimensions.

When variations in the quality of the competing products are not possible, or not permitted, or not feasible, so that the resulting competition centers exclusively on the price variable, the situation may be called one of *pure price competition.* Similarly, when it is not possible or feasible to vary the price at which competing products are sold, so that the resulting competition centers exclusively on the quality variable, the situation may be called one of *pure quality competition.* When both price and quality are variables, competition is neither "pure price" nor "pure quality" competition, but a compound of both. It may then be called *complete competition.*

In some respects the two dimensions of competition are analogous and symmetrical. Both are ways of adjusting the specifications of a bargain. The efforts of sellers to gain customers by offering to sell at a more attractive price bring pressure on all producers to reduce prices and costs. Similarly, their efforts to gain customers by offering preferred qualities bring pressure to adapt quality to buyers' demands. These are two different yet analogous ways of giving the buyer "more for his dollar." Both pressures produce equilibrating movements; in the first case, a movement toward price equilibrium; in the second, toward product equilibrium.

Yet the price and quality dimensions are by no means alike. Price is comparatively simple. Not wholly so, of course—variations in the terms of settlement, credit, quantity discounts, etc., complicate matters somewhat—but it remains largely true that there is only one way in which price can vary: in its numerical value. Quality, on the other hand, is usually a compound of numerous elements (for example, in a necktie: size, shape, type of construction, pattern, colors, material, texture, durability, resistance to wrinkling, colorfastness), each of which is variable. Price, being quantitative, is easily measurable, whereas many of the elements in quality cannot be measured with any precise yardstick. Thus the job of determining the better of two dissimilar prices (when the competing products are identical) is so simple that a mere schoolboy can do it instantly, whereas the job of determining the better of two dissimilar products is a matter for considerable thought and often elaborate tests. Moreover, though all buyers employ the same standard in comparing prices, always preferring a lower to a higher price, they frequently (and quite properly) employ different standards in judging quality, so that it cannot be taken for granted that all buyers will agree as to which of two qualities is to be preferred. Thus a situation in which identical products are being sold at dissimilar prices cannot be expected to persist, while a situation in which dissimilar products are being sold at identical prices may persist indefinitely. These differences between the price and quality facets of competition have tended to obscure their comparability and the fact that they are complementary elements in the evaluation of competitive products.

4. WANTED: A THEORY OF COMPETITION

In this chapter and the preceding one a concept of economic competition has been developed and defined which is in accord with common sense and general usage, but which differs from that usually employed or implicit in price analysis. It differs in two important respects: (1) it is a broader concept, including price competition as one of its constituent elements; and (2) it is a dynamic concept, involving inequalities, disequilibrium, change, and process.

It cannot be asserted that this particular concept is the "true" or "correct" concept of competition. How one chooses to define a term for purposes of scientific inquiry is not a matter of right or wrong, but a question of convenience and usefulness. Nevertheless, it seems worth pointing out that if the foregoing analysis is accepted, the conclusions which it leads to must also be accepted, and that these are at variance with those generally reached in economic theory. The conclusion, for instance, that quality differences and changes are a competitive element, enhancing rather than diminishing the degree of competition in a market, runs directly counter to the well-established tenets of current monopolistic competition theory.

Why does not contemporary theory reveal the competitiveness of quality differences and changes? The answer, it would seem, lies in the fact that theorists have been almost wholly absorbed in the problem of price determination. It is significant that the theoretical study of competitive markets is considered to be a part of the theory of price. Perhaps this has been a historical accident; nineteenth-century economists considered the problem of price far more important than the problem of quality and became preoccupied with the study of "value," and twentieth century economists followed their lead. But whatever the cause, price economics is dominant. To gain quantitative precision, price theorists found restrictive assumptions necessary. They therefore held qualities constant. Their techniques have been of invaluable assistance in developing highly useful tools of price analysis. But their use of them and their preoccupation with price have not been without drawbacks. The technique of keeping quality invariant has obscured the fact that price competition is not the only brand of genuine competition. For while it brings into sharp relief the continual process of price adjustment that occurs in competitive markets as each seller strives to avoid being put at a competitive disadvantage, it hinders the student from observing that in typical real-world markets a parallel process of quality adjustment also takes place in which each seller is also striving to avoid being put at a competitive disadvantage.

The upshot has been that "competition" in economic theory has come to mean "price competition." The term "imperfect" is used to denote deviations from this particular type of competition. These are imperfections from a limited viewpoint; it is of course *price* competition, not necessarily competition *as a whole*, that becomes less perfect.

Thus economists now have at hand a well-tested and highly refined theory of competitive price; but no full-fledged theory of competition. To complement the partial analysis of existing theory an equally rigorous theory of competitive product is needed. Most existing theoretical studies of nonprice competition do not contain the kind of analysis needed, being essentially studies of price behavior. Our study of quality differentiation and variation should not be confined to their influence on price.

How, then, should the quality dimension of competition be incorporated into the schema of equilibrium theory? It is always a sensible procedure in theorizing to isolate the element one wishes to study. Price theory has already done this; by stilling the qualitative movements that are as characteristic of economic life as price movements, it has been able to concentrate on the price aspect of competition. An analogous procedure is possible in working out a theory of quality: price can be kept motionless and quality allowed to vary. Chamberlin has already given an illuminating demonstration of the use of this technique, in Chapter V of *The Theory of Monopolistic Competition,* where, price being held constant, rival producers seek to vary their products in such a way as to maximize their positions. The content of that chapter does not furnish the type of analysis needed, however. Chamberlin's assumptions that quality differences are monopolistic, that each producer's position is impregnable, and that the cost and demand curves of every conceivable variant that a producer might select are fully known to him result in an analysis different from the kind called for by the ideas developed in this volume. But the fact that a different approach seems called for need not bar us from employing this useful technique. After the two dimensions have been studied in isolation, the final step in the analysis should be to allow both to vary simultaneously.

This procedure will be followed in the next three chapters. Quality competition will be examined first alone, then in combination with price competition.

Chapter X

QUALITY COMPETITION

IN this chapter and the next our objective will be to examine quality variability as a competitive element in market behavior. Prices and costs will be kept on the periphery of our vision. We shall not need to ignore wholly the fact that buyers and sellers do not find each other effortlessly; investigation efforts on the part of buyers and advertising efforts on the part of producers will be regarded as adjuncts of both quality competition and price competition. But the fact that the amount of such efforts is variable, is productive of variable results, and is a matter for individual determination will not occupy our attention.

Also, little attention will be paid to deceptive quality variation. It seems analytically more convenient to classify such conduct under deceptive advertising than under product variation. If a quality change actually lacks merit yet brings about a rise in demand because buyers think the product is something other than it is, it seems better to attribute the enhanced sales and profits, not to the quality change (since buyers "chose" to ignore it) but to the misleading advertising (since buyers acted on the basis of it). Thus we shall distinguish sharply between two kinds of situations, both of which involve what is ordinarily called "skimping" or "debasing" of quality. In the first, the producer's downward alteration of the product's quality has been made in the hope that the change will pass unnoticed, so that a reduction in cost can be secured without a corresponding reduction in demand. This is a deterioration of quality, and of welfare. In the second situation, the change is made openly in the belief that buyers would actually prefer the poorer quality at a reduced price to the higher quality and price. What occurs is an experimental adjustment—genuine, effective quality variation. If buyers prefer the cheaper quality, the change can be called an economic improvement; for although the quality is undeniably lowered, purchasing power is released for use elsewhere and consumers' wants are now satisfied with greater precision.

The temptation to pass judgment on cheapening of quality will also be resisted. Free-market societies have been criticized on the grounds that the market creates a bias in favor of shoddy mass-produced goods and that competitive pressures tend to drive quality below its optimum level. This question will not be explored. Actually, the optimum level of quality cannot be determined on economic grounds alone. Personal values must be consulted, even when there is common agreement on cultural goals. To desire fine workmanship is commendable, most of us believe; but it is also commendable to want to enrich one's life by employing one's limited income so as to attain the maximum volume of worthwhile experiences. Which should one prefer, a high-quality watch or a cheaper watch *and* an etching? Precisely where the balance should be struck in any given instance is an intriguing and important question, but we must abstract from it by assuming that it has been solved satisfactorily by each individual.

1. KINDS OF QUALITY COMPETITION

It is desirable to define the term "quality competition" unambiguously. Let us say, then, that it means the simultaneous offering by independent rival sellers of qualitatively different products, at least some of which possess quality features which may be more desirable to some buyers than those of rival products. The term "quality" is given the broad meaning employed throughout this study, referring to all qualitative properties of the product: size, shape, design, materials, etc. Products are considered to be competitive with each other only if they can be regarded as alternative means to a given end. Thus the boundaries of an industry, as defined in Chapter VII, mark off the arena in which any particular competitive struggle takes place. This means that we shall ignore the "competition for the consumer's dollar" that occurs between industries.

The question may arise as to whether quality competition is actually different from price competition. To improve quality is to give more; and to give more for the same price is simply an inverted way of reducing the price. If gasoline is sold on the basis of so many gallons for a dollar, the quantity being variable, price competition takes this inverted form. This method of adjustment is fairly commonplace in the retail field, in circumstances where an exchange ordinarily involves a small sum of money, a single unit of whatever product is purchased, and a physical transfer of currency. Instead of reducing the price of a five-cent candy bar to $4\frac{3}{4}$ cents (which would be a meaningless reduction for most buyers), the producer can accomplish virtually the same result, while retaining the convenience of the round-number price, by increasing its size 5 per cent. Draught beer is commonly "priced" in this fashion, a standard beer being sold at the standard price but in glasses that vary in size with the tone of the establishment or the intensity of the competition. It is perhaps open to question whether a purely quantitative change in the product, affecting only the size of the unit, can properly be called quality variation. If the unit is merely an accounting device, as in the case of gasoline dispensed by a metered pump, quality surely remains unchanged. On the other hand, if the unit involves a physical package, a change in convenience may well be involved, which is a qualitative matter.

Closely allied to inverted price variation is the type of quality change that is similar in intent and effect. For example, a restaurant that wishes to raise the price of its cup of coffee without actually altering the price on the menu can either serve the coffee in smaller cups or use a cheaper grade of coffee. The first is a change in size, the second a change of quality. But both have the same end in view as a price increase: to give less for the money.

This calls our attention to the fact that differences in quality often consist of differences in the *quantity* of some desired ingredient or attribute: the tensile strength of steel wire, the poison content in a given amount of insecticide, the mileage that can be gotten out of an automobile tire. We may then talk of "higher" and "lower" quality, "cheap" versus "good" quality, "improvement" and "deterioration" of quality.

But not all quality differences can be described to everyone's satisfaction in such terms. In some matters of quality there is no clear-cut agreement; different people will rank qualities in different orders. The most obvious examples of differences of this sort are those that involve aesthetic considerations: color, texture, shape—for example, the most pleasing color scheme for porch furniture, the cut of a woman's dress, the pattern of a necktie. What attracts some buyers may repel others. Matters of convenience similarly provide room for disagreement. How should the gadgets on an automobile be arranged? Or the shelves in a refrigerator? Perhaps the best illustration of quality differences that appeal to differences in taste is the one-price table-d'hôte dinner. Here the various alternatives are presumably of approximately equal cost and equal attractiveness. Choice rests purely on personal preference.

It will be useful to label such differences as *horizontal* differences—as opposed to *vertical* differences, which may properly be thought of in terms of "higher" or "lower."

It is to be noted that horizontal quality differences do not usually involve any appreciable differences in cost. One variety is about as cheap to produce as another. In some cases the preferable product may even be cheaper; a plain parchment lampshade may be preferred to one with a picture painted on it. Wherever cost differences are involved, they can always be eliminated (either actually or conceptually) by making concurrently an offsetting vertical quality change. Unlike vertical quality, horizontal quality is of a sort that can become excessive (for example, the stiffness of the springs of a mattress, or the hardness of a soft-boiled egg). Also, horizontal differences are usually unmeasurable, whereas most vertical differences are capable of objective measurement and laboratory testing.

It will sometimes be useful to think of two products as differing horizontally in quality when the "quality" of each actually consists of a complex of measurable, universally desired qualities, some being higher in one product than another while others are lower. Since the component qualities are incommensurable, except in terms of a personal value system, measurability is of no avail; the numerical values cannot be summed by an objective procedure. Which is preferable, a new Chevrolet or a used Buick at the same price? If tire A provides less blowout protection than tire B, how much extra mileage must it provide to make it equally "good" in quality? Here, as in the case of simple horizontal variation, answers must rest on personal value judgments. Acceptance of the fact of significant differences in people's characteristics, tastes, and circumstances, requires us to conclude that dissimilar products cannot in general be ranked in a unique order, as can dissimilar prices. Except under special circumstances, therefore, quality changes are not equivalent to price changes, and quality competition is something other than an inverted form of price competition.[1]

There is also a third way in which quality may be varied. A novel quality may be introduced which is judged superior by most or all buyers, yet costs no more to produce or is well worth whatever additional cost is involved, so that the older quality must eventually become obsolete. A new idea having universal or near-universal appeal (for example, the three-passenger front seat) or a technical improvement in design (steel-shafted golf clubs) lies in this category. In either case the superiority is not due to better materials or workmanship, nor does the change alter the grade

[1] The fact that some qualities are neither "better" nor "worse" than others, but merely better suited to some people's wants (and less suited to others') seems so self-evident that the statement presented above may be deemed superfluous. Yet the idea needs emphasizing since economists have so frequently written as if all significant quality differences were of the vertical type. See, for example, Alfred Marshall, *Principles of Economics*, pp. 100, 113; Stephen Enke, "Profit Maximization under Monopolistic Competition," *American Economic Review*, XXXI (June, 1941), p. 322; Melvin W. Reder, *Studies in the Theory of Welfare Economics*, p. 74.

The related idea that quality differences always involve cost differences has also been expressed from time to time. See Chamberlin, *Monopolistic Competition*, p. 96; Joan Robinson, *Economics of Imperfect Competition*, p. 90 *n*. Yet for a direct statement to the contrary see Theodore Morgan, "A Measure of Monopoly in Selling," *Quarterly Journal of Economics*, LX (May, 1946), 463. Note also that Chamberlin's theoretical model of monopolistic competition requires us to imagine quality differences independent of cost differences, which persist, apparently, because of differences in people's tastes or needs.

or price class to which a product belongs, but rather creates quality differences and improvement within each grade. Such a change may be described as *innovational.* If considered an improvement, an innovational quality change eventually makes obsolete the previously accepted quality, whereas differences in vertical quality, being accompanied by cost and price differences, show no tendency to be eliminated.

To summarize, there are three kinds of quality variability:

1. *Vertical*—the kind of quality change or comparison which may properly be described in terms of "higher" or "lower." Two things distinguish this kind: (*a*) the "superior" of any two qualities is considered preferable by virtually all buyers, and (*b*) it entails greater cost. Therefore an upward (or downward) vertical change in quality unaccompanied by a change in price gives the buyer more (or less) for his money than before.

2. *Horizontal*—those differences about which there is no clear-cut agreement. Two things distinguish this type: (*a*) different people will rank dissimilar qualities in different orders, and (*b*) cost differences, if any, are purely incidental. The existence of this category depends on the fact that people differ in their circumstances, values, and tastes. With differences of this sort we may properly speak of one quality being "more suitable" or "more appealing" than another, but such a statement is meaningful only if made with reference to a particular buyer or group of buyers.

3. *Innovational*—changes which are considered improvements by most or all buyers, yet involve no increase in cost or else are judged superior in spite of whatever additional cost is involved, so that the new quality displaces the old. This kind of quality change, associated with progress, leads to "improved" rather than "higher" quality.

We may thus conceive of three "directions" in which quality may be changed. Quality may be moved *upward* through vertical variation, resulting in "better" quality, or *sideways* through horizontal variation, resulting in "more suitable" or "more appealing" quality, or *forward* through innovational variation, resulting in "improved" or "more efficient" quality.

Actual quality changes do not always fit neatly into one or

another of these categories. A change may be a hybrid, or may contain elements of all three, or may be innovational with respect to some horizontal qualities but not with respect to others. Our classification is simply an analytic device, imperfect yet highly useful for detailed analysis of quality competition.

The fact that vertical differences create different levels of quality enables us to distinguish two broad kinds of quality competition: (1) competition *between* quality levels, and (2) competition *within* quality levels.

Quality competition of the first kind gives rise to price differentials. The differentials may be magnified by selling techniques which mislead buyers into supposing that the vertical differences are greater than they actually are. They may also be magnified by the existence of inequality of wealth and incomes. High-priced products are a symbol of affluence and success; envy may reduce the utility of low-priced products. These are complicating factors; yet such phenomena are traceable not so much to the differences in quality—though the fact of difference is a necessary condition—as to the inequality of purchasing power and the ignorance and irrationality of consumers (or their culture which overemphasizes prestige).

To some extent different quality levels are noncompeting. Buyers have an understandable tendency to shop within a limited vertical range of quality; used Fords and new Cadillacs are not often weighed as alternatives by the same person. The most active competition between different quality levels is between adjacent levels. And the natural barrier between quality classes is often heightened by the fact that price differentials are likely to exceed cost differentials, due to the retailer's custom of adding a constant percentage markup, and sometimes also to manufacturers' differential pricing policies. On the other hand, barriers between price classes are broken down to some extent in the case of durable goods by the fact that goods deteriorate while in use. One can choose on fairly equal terms between a high-quality, high-priced product that wears well but becomes progressively shabbier and a lower-priced product which is poorer in quality when new but can be replaced more often.

Competition among horizontally different qualities is not impeded by income barriers, though it may be subject to other barriers. If tastes are markedly different, for instance, the market may be split into noncompeting groups (for example, tabloids for gum-chewers, full-sized dailies for the intelligentsia). Nevertheless it is probably safe to say that the most intense competition occurs within quality levels. Here we find both horizontal and innovational variation, largely consisting of experimental improvement in adapting products to existing tastes through new combinations or slight modifications of existing features. It is this kind which will engage the major share of our attention.

2. THE CONDITIONS OF QUALITY COMPETITION

Do quality differences and changes always imply quality competition? Or merely under certain circumstances? The following conditions are suggested as those which are necessary to assure the presence of quality competition.

1. *Dissimilar alternatives must be possible.* In some industries there is little or no room for dissimilarity of product. In the sale of refined metals or chemicals, for example, the one significant variable is the degree of impurity; and if buyers accept some uniform standard of purity for the sake of convenience and find noticeable deviations from the standard unacceptable, or if the degree of impurity makes no difference provided allowance is made for this in the product's price, as in the sale of potash,[2] quality competition is eliminated so far as the physical product is concerned. It is not eliminated entirely, of course—merely narrowed to the variable services provided by the seller.

2. *It must be possible either to vary the alternatives (that is, provide new ones) or to vary the buyers' knowledge concerning them.* If only *n* alternatives are feasible and all of them are already being offered by rival sellers, and buyers are already familiar with them, yesterday's evaluations remain valid for today's trans-

[2] See Samuel P. Hayes, Jr., "Potash Prices and Competition," *Quarterly Journal of Economics*, LVII (Nov., 1942), 31–32.

actions and yesterday's competition need not be repeated. If the currently available alternatives cannot be changed, variability of knowledge provides some room for quality competition. Sellers can engage in competitive efforts to attract new customers by removing erroneous unfavorable impressions (or by creating erroneous favorable impressions). And whenever new buyers enter the market a new contest must take place for their benefit. But under such stationary conditions the cumulation of experience is likely to narrow progressively the area in which knowledge is changeable, and it seems safe to predict that quality competition would diminish continually in intensity and importance. The possibility of product variation is therefore a major requirement for active, continuous quality competition.

3. *Producers must be free to introduce novel, "nonstandard" qualities.* When innovations in quality are possible but not permissible, quality competition is thwarted. Enforced standardization ordinarily destroys or diminishes quality competition. Other restrictive regulations of a comparable nature, such as local building ordinances, may also lead to its impairment. Of course, when alternatives are prohibited which sensible buyers would never knowingly select, it is only a through-the-looking-glass kind of quality competition that is prevented. On the other hand, when controversial qualities are suppressed (a possible byproduct of well-meant regulation) or when special interest groups employ regulation as a device to further their own ends, there is likelihood of a real impairment of socially desirable opportunities for choice.

It is important to realize that standardization does not always diminish quality competition. Some kinds of standardization actually enhance it. Standards of weights and measures are of course a prerequisite of both price and quality competition; it would be impossible to determine whether an undisclosed amount of gasoline A is preferable to an undisclosed amount of gasoline B. Allied to this is uniformity of size, a form of standardization, which often makes competition more effective by making both quality comparison and price comparison easier. Standardization of fittings is another invaluable aid to competition in some industries. Whenever complementary products must be fitted together in order to be used (for example, razors and blades, beds and mattresses, incapacitated machinery and replacement parts), standardization makes possible interchangeability and thus makes available to the possessor of a product any one of the products in the complementary industry. In fact the deliberate use of nonstandard fittings is a frequent device by which producers of joint products seek to eliminate competition in one of the two industries (for example, the "improved" Gillette razor of some years back, which would take only the patented Gillette blades). It should be kept in mind that the standardization of portions of the design of a product does not increase competition in that industry, but only in the complementary industry. Putting standard-size wheels on automobiles theoretically narrows quality competition in that industry (though this is probably without significance), since it eliminates one variable element in the quality of competing automobiles; but it intensifies tire competition, since every tire manufacturer can now afford to produce all the standard sizes and become a contender in every corner of the market, whereas otherwise the tire industry would be partitioned off into many small markets, each offering little or no opportunity for price or quality alternatives. Similarly, the adoption of a nonstandard turntable speed introduces a new element into quality competition in the phonograph industry; but it diminishes both quality and price competition in the record industry.

4. *Producers must be free to try to imitate the qualities of rival products.* It is not a required condition that they be *able* to imitate competitors' products exactly—only that they are not restrained from attempting to do so. (The analogy of athletic competition is helpful here; it is neither necessary nor feasible to guarantee to each runner that he will be able to run as fast as the fastest, but it is necessary to give him the chance to do so.) Thus a quality change is not unrestrictedly competitive if the producer's new position is protected by a patent or copyright. It is competitive to some degree, however, since it will normally spur competitors to fresh innovations. Only if all avenues to advantageous innovation have been successfully blocked off is a protected quality wholly monopolistic. For it makes no difference from the standpoint of

competitive advantage whether a competitor's response to a successful innovation is a precise imitation or a product which, though different, is equally attractive to buyers. On the other hand a quality difference due to superior skill, which others are free to try to imitate, is unreservedly competitive.

A somewhat unusual example of competitive imitation is the familiar rivalry between phonograph record companies to acquire the services of celebrated musicians as exclusive artists. Superficially this smacks of monopoly. Yet the dynamic picture is not one of monopoly; when contracts expire, rival firms try to "imitate" successful products by acquiring the artists responsible for them, the result being that changes in allegiance are frequent. And even from a static viewpoint it is not monopoly, since an artist's past recordings remain the property of the recording firm. Columbia's Burl Ives records compete with Decca's Burl Ives records.

5. *Decisions of rival firms in matters of quality must be made independently.* The reason for this is partly that producers have a vested financial interest in their own current varieties. It is often cheaper not to bring out new models unless the threat or actuality of a superior rival product forces the firm to make the change in self-defense. Producers may also have a sentimental, irrational bias in favor of their own brain children. But probably the most important reason is that people have imperfect knowledge, limited imagination, and timidity in trying the unknown. In a world such as ours the merits of socially desirable innovations in quality are not always readily apparent in advance. As in scientific research, the existence of numerous independent decision-making units provides the best insurance against mediocrity and stagnation.

Collusion does not necessarily destroy all quality competition; but it may produce a situation in which trivial or token quality competition occurs while, by secret agreement, major quality improvements are avoided. Similarly, a nationalized or privately monopolized industry need not lack quality competition; but if decisions are centralized, so that ideas for possibly advantageous variants which fail to win the approval of the central management die unborn, the produced alternatives may be significantly different in number and character from those likely to be made available under unfettered quality competition. This is not to say that vigorous quality competition between divisions of a large corporation or between producing units in a socialized industry is impossible. But it seems correct to say that a condition of such competition is some degree of autonomy in quality determination on the part of the rival divisions.

6. *Buyers must have access to information about quality differences, and must avail themselves of it.* Obviously there can be no quality competition without quality comparison. If sellers of differentiated products induce consumers to buy without making comparisons, or impair their opportunity to compare, or if buyers do not take the trouble to examine quality, quality competition is thwarted.[3] Since quality is more difficult to evaluate than price, this entails fairly elaborate institutional arrangements: advertising, aids to personal investigation, and for some kinds of products advisory organizations devoted to the testing and rating of the alternatives.

Each of these six conditions is capable of being either partially or completely realized. Thus there is room for a good deal of variation in the intensity of quality competition in any market. The intensity will vary with (1) the natural opportunities for quality competition, which depend on the nature of the product (for example, its complexity) and the variety of uses to which it may be put, and (2) the extent to which the economy's institutional arrangements permit use to be made of these opportunities. The possible opportunities as determined by (1) must be accepted as given —at least for the economist. On the other hand, the degree to which these possibilities are actually realized is a variable that can be affected by appropriate changes in government policy. Let us call the first the *degree of opportunity for quality competition,* the second the *degree of freedom of quality competition,* and the combined result the *intensity of quality competition.* If we wish to determine whether or not genuinely unfettered quality competition exists in a given industry we must focus our attention on the second of these variables.

[3] Cf. Robinson, *Economics of Imperfect Competition,* p. 186; Corwin D. Edwards, *Maintaining Competition,* p. 7.

Determining the degree of freedom of quality competition in an industry is not always easy. The direct method would be to study the industry with reference to conditions 3, 4, 5, and 6. An alternative method would be to determine first the opportunities, by studying the industry's product, demand conditions, production techniques, engineering developments, etc., with reference to conditions 1 and 2, and then to estimate the actual intensity of quality competition.

Several possible indicators of the intensity of quality competition may be consulted:

1. *The number of available varieties.* Quality competition might be said to have reached its maximum intensity when the addition of one more variant does not enable a substantial number of buyers to satisfy their wants with appreciably greater precision. The numerical value that corresponds to this condition will of course vary widely from industry to industry.

2. *The degree of quality flexibility.* Just as price flexibility is an indicator of price competition, so the frequency of quality changes and the promptness with which firms respond to innovations in quality introduced by competitors are signs of active quality competition. Quality variation is apt to be more sluggish than price variation for technical reasons. Retooling may be required, the routine of purchasing and stocking raw materials may be upset, and production methods are bound to be affected; all this is more cumbersome and costly than printing a new price list. Producers of some types of products are also likely to be reluctant to make quality changes, wishing to reap the full advantage of buyers' familiarity with existing models. And even when opportunities arise for quality changes that are worth adopting, entrepreneurs may remain blind to them. Obviously the maximum degree of practicable flexibility varies widely with the nature of the industry. Yearly changes in automobile designs are probably as frequent as can be expected, perhaps even undesirably frequent, while a weekly magazine cannot help but vary in quality with each issue.

3. *Variations in competing firms' shares of the market, if correlated with product variation.* The purpose of an innovational change in quality, from the innovator's standpoint, is to gain a

larger share of the market. Even if competitors can and do subsequently imitate the innovation or wipe out the competitive advantage with counter-innovations, there is a time interval in which a substantial shifting of buyers' allegiance can occur. Also, competitors may be skeptical of the merits of an innovation until falling sales convince them that their customers prefer the new quality. Competition becomes intense only when the rival contestants must strain every muscle not to be left behind in the race. Quality variations to which they need not respond are not very competitive.

4. *Changes in the composition of an industry: the entry of new firms and the exit or decease of old ones.* This is merely an extension of the point made in the preceding paragraph. Just as price competition drives out firms unable to keep their costs down, so quality competition drives out firms (if there are such) which lack skill in adjusting quality to buyers' wants. If there is a tendency for established firms to become overconservative, it may require the entry of new firms with new varieties to reinvigorate quality competition.

5. *Conformity to a distinctive pattern of "challenge and response."* Our analysis suggests the hypothesis that quality competition, viewed as an experimental search for more advantageous varieties of the product, finds expression in a recurrent cycle of events, occurring in a predictable sequence. First comes an innovational change in some firm's product which increases the "distance" between it and its closest substitutes. If this is followed by an increase in the innovator's share of the market, competitors respond with changes in their products which decrease the "distance" between them and the innovator's product. If, on the other hand, the original innovation leaves the innovator's share of the market unchanged, or diminishes it, any ensuing quality changes made by competitors are more likely to be movements "sideways" or in the opposite direction. This is not the only possible pattern. Monopolistic competition theory implies a quite different one: An increase in the "distance," if successful, causes the innovator's demand curve to tilt downward more sharply; the measure of success is not so much a rise in sales volume as an increased profit margin, shown by the greater disparity between average revenue and marginal revenue;

this indicates success in building a sheltered market. (If sales volume remains unchanged, the MR curve has simply rotated on the point of MR-MC intersection.) Rival producers thereupon seek sheltered markets of their own, and make changes which increase the "distance" still further. The result is a general scattering movement, resembling the scientists' picture of our "exploding" universe. If observable product variation takes this form, it is clearly not competitive.

Chapter XI

EQUILIBRIUM UNDER PURE QUALITY COMPETITION

LET us suppose a market in which quality competition exists alone. What equilibrating adjustments will take place? And what will be the final position of equilibrium? In order to answer these questions, we shall need to employ a theoretical model of a market which is "pure" and "competitive," yet is precisely the reverse of the so-called "purely competitive" model. Here we enter the Looking-Glass country of *pure quality competition.*

1. THE CONDITIONS OF PURE QUALITY COMPETITION

Pure quality competition, we have already seen, is a form of incomplete competition, whose conditions are the polar opposite of those of pure price competition. It occurs when the price of all competing products remains uniform and unchanging, while quality is freely variable, so that adjustments are restricted to changes in product quality—with, of course, corresponding changes in output. In order that price competition be inoperative, identical prices must be maintained either by law or by private compulsion of some sort or by virtual necessity, as when strongly established custom (for example, in the size of a physician's fee) or the indivisibility of monetary units or coins (in products dispensed through slot machines) or convenience (the newsstand price of the successive issues of a newspaper or magazine) makes price variation so undesirable that producers will not resort to it. In order that quality competition be unhindered, the conditions enumerated in Section 2 of the previous chapter must be fulfilled.

Pure quality competition, like pure price competition, is not often found in the real world. One interesting instance of it is radio and television entertainment, which competes for audiences purely on the basis of quality, the price in every case being nil. Another is retailing service in areas in which resale price maintenance laws are in force. But our chief interest in it lies in the fact that it displays with great clarity types of behavior and adjustment which are also found in markets where both products and prices are free to vary, and which are especially significant in the many markets of today in which prices, although not completely rigid, move sluggishly.

2. THE CONCEPT OF QUALITY EQUILIBRIUM

Quality adjustments are typically at least partly innovational in character. Even if a change is, strictly speaking, horizontal, some experimentation is usually involved, for whenever a producer offers a hitherto unproduced quality (or combination of qualities) he is likely to be somewhat uncertain as to consumer response. We can get around the uncertainty difficulty by supposing perfect knowledge—or, alternatively, by neglecting all the false moves and trial-and-error processes and looking only at final equilibrium. But if quality equilibrium requires that all possible innovations have been made, we face a more serious difficulty. For innovation is endless. Human ingenuity is capable of devising countless variations of products. Progress in science and technology opens up ever-new opportunities for further variation. New variants of existing products become transformed into virtually new products with new uses (for example, the horseless carriage, originally a close substitute for horse-drawn vehicles), making further equilibrating adjustments of quality necessary. Even in an economy in which people's wants remained constant, the number of moves and countermoves needed for the achievement of quality equilibrium would seem to be indefinitely great.

To make our analysis manageable, we must impose conditions

under which equilibrium is conceptually attainable within a reasonably short period. Therefore innovational variation will be ruled out. An unchanging technology will be assumed.

This restriction does not eliminate all pioneer quality changes. Producers are still free to offer novel variants so long as they do not require the specialized skills of engineers or extraordinary inventive skill on the part of management. Quality changes of two sorts remain possible. A new quality may be selected "in between" the qualities of two variants already in existence. (If bicycles of 24-inch and 26-inch wheels are already being produced, producers are free to offer a 25-inch-wheel bicycle.) And in the typical case in which a product's quality is compounded of numerous variable elements, existing quality elements may be put together in new combinations.

The idea that some novel qualities involve technical progress while others do not may seem questionable. All innovations and inventions can be described as new combinations of previously existing devices or materials or ideas.[1] The amount of creative imagination and technical difficulty involved in quality variation is probably continuously variable, so that while some changes are clearly of a routine nature, and others clearly involve technological advancement, the two kinds merge imperceptibly into each other. But for our purposes the distinction seems a valid enough device; in practice an approximate dividing line can be drawn. The distinction being sought here is like the distinction made in genetics between those new organisms which are produced by obtaining new combinations of existing genes, and those which arise through the process of mutation. We are here abstracting from the grandiose Schumpeterian innovation that resembles the sudden appearance of a qualitatively new character and limiting ourselves to the day-to-day process by which new unique individuals are created.

3. GENERAL ASSUMPTIONS

In analyzing quality equilibrium, specific assumptions are of course needed. We must keep our theoretical model simple, especially since we are breaking some new and rather difficult ground. Yet we

[1] C. E. Ayres, *The Theory of Economic Progress*, pp. 112–14.

must avoid oversimplifying to the extent that the essential problems of quality determination in competitive markets are assumed away. With these two thoughts in mind, the following assumptions are made, to be retained without modification throughout the analysis:

1. *Each firm offers only one variety of the product.*

2. *Each firm is free to choose any variety among those possible, whether being produced currently by some other firm or not.* This assumption, quite different from the Chamberlinian assumption that each firm has a monopoly of its own variety, exact imitation being impossible, is made in order to make our analysis applicable to situations in which quality competition is not alloyed with monopoly elements.

3. *Quality is continuously variable.*

4. *All buyers are consumers.* This assumption is not essential, but the explanation can be kept simpler if the motives and behavior of business buyers are neglected. Actually, the analysis can be applied to producers' goods markets with only slight modification.

5. *Buyers act rationally and respond promptly and correctly to every change in the quality alternatives made available to them.* This amounts practically to assuming perfect knowledge of utility on the part of consumers. In actual markets quality adjustment is complicated by the fact that producers may be temporarily misled by errors in consumers' estimates of the utility of newly offered variants. If the errors are eventually discovered, equilibrium will be reached, but by a more circuitous route. It seems better to avoid this complication since we are now interested in the equilibrium state itself, not the route by which it is reached. This assumption also rules out producer exploitation of consumer ignorance, apathy, and gullibility.

6. *Buyers' preferences and demand functions remain constant.*

7. *Technology remains unchanged.* Neither innovational changes in quality nor changes in cost functions are possible. This assumption reduces quality variability to two kinds: horizontal and vertical.

8. *Every firm has the same cost function, unit cost varying with rate of output and with vertical quality but not with horizontal*

changes in quality. Costlessness of horizontal variation seems reasonable to assume. In the actual world changes in cost, if not nonexistent, are likely to be minor and purely incidental.

9. *Cost functions are such that all average cost curves are U-shaped.* This assumption takes into account the diseconomies of both small-scale and large-scale production. It would be simpler, of course, to assume constant cost. But this would vitiate our entire analysis. If there were no diseconomies of small-scale production, firms would tend to become more and more numerous, each producing less and less, until the number of available varieties equaled the number of optimum points on the quality scale. If every buyer had a different optimum point, the number of firms would at least equal the number of buyers. If only a limited number of horizontal varieties were wanted, each considered optimum by numerous buyers, there would still be at least as many firms as wanted varieties. In either case the actual number would be indeterminate.

All this is too unrealistic to be rewarding. Moreover, it shelves the central problem, which occurs only when the high cost of small-scale production limits the number of varieties that can be produced to a number *less* than that which would satisfy all buyers with perfect precision. When this limitation exists, an interesting problem in maximization is presented, which we shall want to explore.

10. *Decisions of rival firms in matters of quality are made independently.* There is no collusion.

11. *No firm's decision is influenced by considerations of strategy or game-playing arising because of expectations concerning moves which particular rival firms might make in response to its actions.* We are abstracting from oligopolistic interdependence of decision-making.

4. HORIZONTAL EQUILIBRIUM

To avoid encountering too many complications at once, it is well to examine horizontal and vertical variability separately. Let us start by supposing a market in which both price and vertical quality remain fixed; only the horizontal dimension of quality is variable.

To further simplify matters let us assume that

12. *The product is horizontally variable in but a single respect, and its variability is such that the possible varieties can be arranged in order as if along a scale.* In other words, what we shall have to deal with is an "array" or "spectrum" of horizontal quality.

This is, of course, a drastic simplification of reality. Normally products are horizontally variable in numerous respects; in fact, it is this multidimensional character of quality that makes choice between quality alternatives so much more difficult than choice between price alternatives. And it is not always true, even with respect to each of the separate variable elements of a product's quality, that the possible variants can be unambiguously ordered. The idea of gradation does not seem applicable, for example, to types of furnaces (coal, oil, gas) or house exteriors (brick, stone, wood, stucco).

Yet, in a limited and partial sense, this simplification remains reasonably faithful to the facts of our world. Differences that cannot be arranged along a scale are limited in number; most product variation consists of alterations in degree rather than in kind. If all but one of the elements of a product's quality are held invariant, variations in that one element can usually be arranged in a definite order. The various possible wheelbases of an automobile are capable of numerical arrangement; so are the various possible variations in weight, horsepower, width of front seat, glass area. Different shades or hues of a given color can be arranged from lightest to darkest, or from the most vivid to the most subdued. Even such an elusive quality as the character or tone of a newspaper can be arranged, at least conceptually, with the most "vulgar" at one end of the scale and the most "intellectual" at the other.

Restriction of horizontal quality to one dimension enables us to resort to simple diagrammatic representation. Employing the familiar techniques of spatial competition analysis, we may represent the distribution of buyers' preferences for the various possible horizontal variants by their location along a line representing these variants arranged in an array. Horizontal quality variations can then be shown by changes in the locations of producing firms on this line. The variable distance between a buyer and the producer located nearest him will represent the degree of unsatisfactoriness of that producer's product due to the divergence of its horizontal quality from that quality which to that buyer is "ideal."

Additional dimensions would complicate the analysis. To depict the possibilities obtainable by varying simultaneously two of a product's component qualities (for example, the sugar content and cream content of a cup of coffee), a two-axis diagram would be needed; if we wanted to show quantity demanded, a third dimension would be necessary. Faithful representation of horizontal variability is possible only in *n*-dimensional space. It seems wise to avoid this difficulty! At the same time, it must be borne in mind that our analysis will be oversimple, and is illustrative rather than fully descriptive of quality variability.

We shall also need to specify with some precision the pattern of buyers' preferences and demand with respect to horizontal variation. There must exist for every buyer an optimum point on the horizontal quality scale, indicating the variety that has "just the right amount" of the relevant quality.

One might guess that individual preferences are likely to be distributed in most cases in a way that somewhat resembles the normal curve of error, with a bunching of preferences in the central portion of the array, and sparsity in the end zones. People who prefer three- or four-minute eggs are more numerous than those who prefer one-minute or ten-minute eggs. The distribution need not be symmetrical, and might of course be bimodal or trimodal, but in any case there are likely to be limits beyond which the product would become undesirable in practically everybody's opinion. It would greatly complicate our analysis, however, to assume a distribution of this sort. It seems wise to adopt the simplest assumption, even though a considerable loss of realism is involved. Let us therefore assume that

13. *Buyers' preferences are distributed evenly along the horizontal quality scale between two well-defined limits.*

We shall also assume that

14. *The demand for a firm's product on the part of any group of buyers having identical preference systems is a decreasing function of the product's distance from that group's optimum point on the*

quality scale. This assumption is quite in keeping with orthodox economic principles. When quality becomes "poorer" in the minds of buyers, its utility falls. If the good can be consumed at a slower rate or less frequently, buyers will tend to reduce their rate of buying; if not, marginal buyers will drop out of the market. To give this assumption greater precision, let us specify a function of the form $q = m\ (1-d/z)$, where q represents the number of units demanded by buyers located at some position on the quality axis, m the amount that would be demanded if the variety were considered optimum by those buyers, d the distance between product and buyers, measured along the quality axis, and z the distance at which the demand would have fallen to zero. Negative demand of course will be ignored.

15. *Each buyer demands exclusively the variety that lies closest to his optimum.* The demand function specified above measures actual demand for a variety only in that portion of the spectrum in which no other variety lies as close.

Finally, let us assume the following additional conditions, which will later be modified:

16. *Firms seek to maximize profits.*

17. *Vertical quality remains fixed at an arbitrary height.*

18. *The line representing horizontal quality variability is a closed curve.* This assumption avoids the complication of "end firms," which will be considered later. It is admittedly less realistic than a straight-line assumption, yet not wholly devoid of realism. Think, for example, of sellers offering a standardized product for delivery once a week (no oftener) to some central location where buyers congregate, each seller being free to select the day and hour of delivery, with buyers' preferences spread evenly throughout the week.

19. *The number of firms is fixed, and is greater than the number that would permit each firm to produce at its most profitable rate.* Each firm, therefore, wishing to increase profits, will seek to increase sales.

Our problem is to discover how the products will be spaced along the quality spectrum.

The situation may be pictured in a diagram, the base line of which represents the possible horizontal variants arranged in an array (and therefore the possible "locations" open to each firm by its choice of horizontal quality), while vertical distance measures quantity bought by consumers located at each position on the quality scale. Since the density of consumer population is equal all along the scale, a horizontal line at some height above the base line can represent the maximum demand at any point. The salable output of an unopposed producer can be represented by an isosceles triangle whose apex is directly above his location on the quality spectrum. If the triangles of two competing producers overlap, the portion of the market retained by each can be shown by drawing a vertical dividing line from a point midway between the two locations. (To consumers located at that point the two varieties are equally unsatisfactory.)

Then let us suppose, at the start, that producers are scattered irregularly along the line. Figure 1 shows a portion of this line. The areas of the tent-shaped figures show the sales of the three varieties,

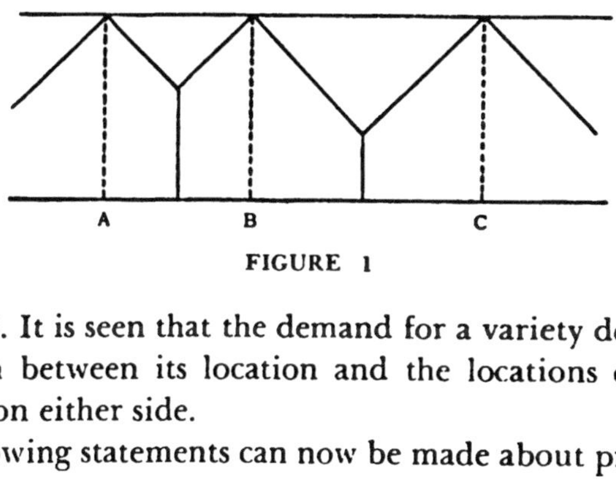

FIGURE 1

A, B, and C. It is seen that the demand for a variety depends upon the relation between its location and the locations of its closest substitutes on either side.

The following statements can now be made about producers' behavior:

1. A producer whose variety lies close enough to its two nearest rival varieties to have its sales restricted by their presence, but at a greater distance from one than from the other, will move to the point midway between them. (So long as the height of the vertical dividing line on one side is greater than on the other, the area gained by such a change in position is greater than the area lost.)

2. When all equilibrating adjustments of this sort have been

made, the products will be "equidistant" in quality, and all firms' outputs will be equal.[2]

Suppose now that the quality spectrum, instead of being a closed curve, is a straight line of finite length. How close to each end will varieties be located?

Unlike producers whose varieties are located between others, an "end producer" will not try to achieve a symmetrical tent-shaped figure. His situation is different; by moving away from the end, he can increase the length of the figure's base. (If he moves a distance *d,* the length of the portion of the base between his location and the extremity is increased by *d,* while the other portion is decreased by only ½*d.*) His position of maximum output is achieved when he adjusts his location so that the height of the vertical line dividing his sales area from that of his nearest competitor is twice the height of the vertical line at the end of the spectrum.

This makes the output of each end firm larger than that of other firms. The amount by which its output exceeds theirs depends on the steepness of the sloping "roofs" of the tents. If the slope were so gentle that, in equilibrium, the end firm's distance from the extremity were more than three halves the distance between each firm, the equilibrium would be unstable. Any in-between firm could gain by relocating between the end firm and the extremity, and an endless reshuffling would occur.

The extra-large outputs of the end firms and their inability to retain their end positions in markets where horizontal preferences are weak have no parallel in the real world. These characteristics of our model are apparently the result of the highly artificial assumption that demand terminates abruptly at each end of the line, rather than tapering off at the outer fringes. It seems safe to conclude that the

2 This conclusion differs from those of Hotelling's and Chamberlin's analyses of spatial competition. Hotelling's celebrated analysis, in "Stability in Competition," *Economic Journal,* XXXIX (March, 1929), 51–52, led its author to conclude that where there are but two sellers both will locate at the center of the line. But this follows from his assumption that an increase in the distance between buyer and seller does not diminish the quantity bought. The same assumption is responsible for Chamberlin's conclusion in Appendix C of *Monopolistic Competition* (p. 261) that "where a seller finds himself between two others . . . it is a matter of indifference at what point he locates" and therefore that sellers may group in twos instead of being evenly dispersed.

conditions of equilibrium with respect to the end firms have no practical significance.

Suppose the assumption of equal distribution of buyers' preferences is replaced by the more realistic one of a normal distribution. How would the equilibrium position be affected?

To explore this question thoroughly would require highly complicated mathematics, and will not be attempted here. Once the "ceiling" (the line indicating the demand at each location for the optimum variety) is tilted from the horizontal, demand functions of the sort assumed on page 146 can no longer be represented by straight-line slopes. If the "ceiling" were tilted upward to the right, the right-hand slope would be convex from above, the left-hand slope concave. If modified demand functions that could be represented by straight-line slopes were adopted, the model would be more manageable but would be hard to justify as a representation of rational behavior. Suffice it to say that if density of population varies in some systematic fashion with distance from the center of the quality spectrum, (*a*) firms will bunch together more closely in the more densely populated than in the sparsely settled areas, and (*b*) their outputs may or may not be equal, depending on the particular demand function assumed.

Another method of dealing with the complexity of a normally distributed population is possible. We can redefine the quality scale so that each unit of distance represents a portion of the spectrum of whatever "breadth" is necessary to encompass the preferred varieties of a given specified number of buyers. This amounts to grouping the possible varieties into many minute classes, such that each class has an equal number of buyers whose optimum variety lies within that class. Since by definition buyers are now spaced evenly along the newly defined scale, the "ceiling" will be, as in Figure 1, a horizontal line, and the same simple geometrical analysis can be employed.

This procedure is perhaps open to criticism on methodological grounds. A unit of quality change is no longer defined in terms of an objective, measurable characteristic of a product, but in terms of buyers' attitudes. On the other hand it can be argued that with some kinds of horizontal variation (such as the "loudness" of a necktie) there may be no other way of defining a unit of quality change. And in some instances of measurable changes (such as temperature of a room) the objectively measurable unit may be much "smaller" subjectively in one portion of the quality spectrum than in another. A drop in temperature from 70 to 67 degrees Fahrenheit might be considered a greater change in one's environment than a drop from 40 to 37 degrees—or from 140 to 137 degrees.

If this device is employed, the conclusions reached earlier with regard to equidistant spacing must apply to the equilibrium position of firms under conditions of normal distribution of buyers' preferences. If we translate our conclusions back into terms of our original quality scale (on which the "ceiling" line would be a bell-shaped curve), we can say that firms will be bunched most closely beneath the peak of the curve and most widely spaced at each end. The

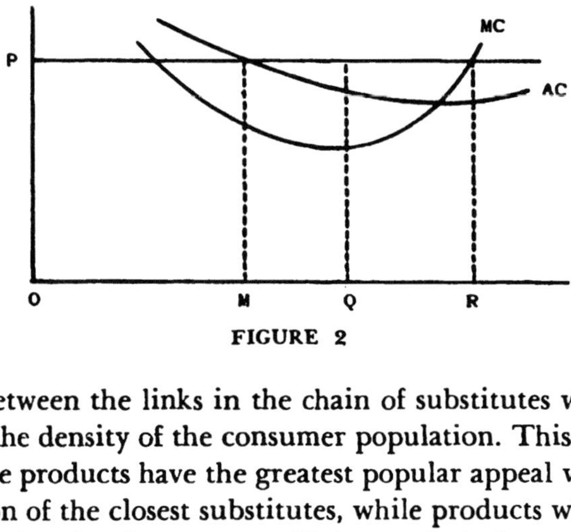

FIGURE 2

distance between the links in the chain of substitutes will vary inversely as the density of the consumer population. This means that firms whose products have the greatest popular appeal will face the competition of the closest substitutes, while products which appeal only to small minority groups will have relatively greater shelter from substitute competition.

Now, let us suppose that firms are free to enter or leave the industry. How many will it contain in equilibrium?

The answer depends on the relationship between price and average cost at the existing rate of output. Suppose each firm's cost curves and selling price are as shown in Figure 2. When all firms have found their equilibrium locations, each has an output rate which, we have assumed, is less than the profit-maximizing rate *OR*. If it exceeds *OM* the firms are enjoying supernormal profits. Let us suppose an output of *OQ*. With price fixed and vertical quality adjustment impossible, profits cannot be eliminated except through the entry of new firms into the industry. New firms will be induced to enter. If they enter while the firms already in the industry are still in disequilibrium, they will of course select the "roomiest" locations; if not, they will select a location at random. At first they will be too close to competitors to be able to attain a profitable volume, but their neighbors will promptly move away from them. Equilibrium is reached when the number of firms n increases until it is the number at which *OM* (the no-profit rate) is $1/n$th of the industry's output.

Similarly, if the equilibrium output rate for the firms already in the industry is less than *OM*, losses will be incurred, and firms will leave the industry until output expands to the no-profit rate.

In the model constructed in this section one kind of quality difference is noticeably lacking: differences between quality grades. This lack is the consequence of our assuming that price and vertical quality are single and invariant. It could be repaired by replacing that assumption with another to the effect that more than one quality grade is permitted, each to be sold at its own fixed price. The consequence would be a division of the market into "submarkets" or "strata." Each buyer would then be faced with the task of choosing between grades as well as among horizontal qualities. If numerous grades were established and price-fixed, the market would attain a character more nearly resembling that of most actual markets, though it would still be characterized by rigidity of vertical quality and price within each submarket.

No attempt will be made here to build a multi-price model. It is worth bearing in mind, however, that for every single-price model a corresponding multi-price model, more realistic but more complex, can always be constructed.

Another thing to note is that in our model each firm has only two effective competitors; no firms other than its closest neighbors can affect its sales volume by horizontal changes in quality. This is true, however, only when horizontal quality variability is restricted to a single dimension. If horizontal quality variation were two-dimensional, three firms could locate so that each was equally distant from the other two; if three quality elements were variable, four firms could be equidistant from each other; and so forth. What looks like inevitable oligopoly in our model vanishes when the multidimensional character of quality variability is taken into account.

We must also note that the equilibrium rate *OM* in Figure 2 is actually exceedingly arbitrary. Its value depends on two conditions either or both of which could presumably be altered: (1) the predetermined price, and (2) the predetermined "level" of quality, which determines the height of the average cost curve. Typically, excessive profits that invite entry of new competitors are eliminated partly by competitive price-cutting or competitive quality-raising. Our reason for assuming *both* price and vertical quality to be invariable was in order to observe the horizontal dimension of quality in isolation. It is now time to relax the second of these constraints.

5. VERTICAL EQUILIBRIUM

Let us start with the position of horizontal equilibrium already reached, and assume that producers are now free to vary quality vertically as well as horizontally. Let us assume also that

20. *Vertical variability is restricted to variation in a single component of vertical quality, such that for every quality "level" or "grade" there is one, and only one, corresponding cost function, and for every output rate there is a different unit cost associated with each different level of quality.*

This restriction enables us to depict quality levels graphically by drawing their corresponding cost curves. In reality, of course, vertical variability, like horizontal variability, is likely to be multidimensional; when this is the case, the level of quality is conceivable only as some kind of average of the levels of the various vertical quality components, and the relationship between cost and vertical quality becomes obscure. The above simplification removes this difficulty.

Assumptions are also needed with regard to the way in which

vertical quality variation affects demand. Two situations must be distinguished: one, in which vertical quality changes are made by all firms in unison; and the other, in which one firm raises or lowers quality while the rest of the firms leave quality unaltered.

Let us consider first market demand as a function of quality height when vertical changes are made simultaneously by all firms. Unless demand is totally inelastic, a rise in vertical quality in a fixed-price market—like a fall in price in a fixed-quality market—will affect the amount demanded to some degree. Buyers will purchase more. Yet the statement just made is ambiguous; does it mean that they will buy more units of the product? or merely that they will purchase enough to bring them more satisfaction? This question does not arise in the theory of price, since when products remain unchanged in quality an increase in satisfaction can be obtained only by an increase in the number of units consumed. But when quality is vertically variable, this is no longer true.

Suppose a situation in which mining companies are supplying ore to a copper-refining firm and now offer a higher-grade ore at the same price. Will the refining firm increase its purchases? If the demand for refined copper is not totally inelastic, the firm will certainly buy a large enough quantity to give it a greater volume of refined copper, but it is altogether possible that a smaller volume of the superior ore will yield the increased volume of copper desired, in which case the firm's purchase of ore will actually be reduced. A similar pattern of response is possible in cases where the desired ingredient is not a physical substance like copper, but a psychic quantity or a measurable service, such as mileage in an automobile tire. In all such cases it is necessary to distinguish between the *volume of satisfaction* desired by the buyer and the *number of units* desired.

To obtain the sharpest picture of buyers' response to a vertical change in quality, let us consider the kind of variation in which an upward change increases the amount of satisfaction obtainable from a product, in the opinion of all buyers, without there being any accompanying decrease in its satisfactoriness in some other respect. Not all "upgrading" is of this sort. Ordinarily a cost-raising quality change involves both a quantitative element (the giving of "more" of something which is invariably preferred in larger rather than smaller amounts) and a qualitative element (the provision of some other quality or qualities which are preferred in some optimum amount, and can be deficient or excessive).[3] Thus it is ordinarily a mixture, an alloy. But the proportions of the amalgam are variable, and we can conceive of a limiting case in which only the quantitative element is present—when the change consists simply of an alteration of the size of the unit, all sizes being equally convenient or economical, or of the amount of some desired ingredient contained in it, buyers being wholly indifferent to any concomitant variation in the amounts of other ingredients present.

If we restrict our analysis to this "purest" kind of vertical quality variation, the pattern of response of consumers to vertical changes can be stated without difficulty. Clearly it must be exactly what their pattern of response would be to price changes in terms of a fixed unit. In any market demand schedule which embraces the gamut between the prohibitive price and the zero price, there is at least one point or range of output (and price or prices) at which the price elasticity of demand is unity; at smaller outputs (higher prices) the elasticity is greater than one, while at larger outputs (lower prices) it is less than one. Similarly, in a fixed-price market in which the unit is of variable size there must be some range of small sizes in which an increase in the unit's size causes a rise in physical volume demanded which is more than proportional to the increase—in other words, it causes a rise in the number of units demanded; beyond this there must be a size, or series or range of sizes, in which an increase raises the physical volume demanded proportionally, so that the number of units demanded remains unchanged; and beyond that must lie another range in which, though the physical volume rises further, the number of units demanded declines as the unit grows in size until finally, at the satiation point, the decline in units demanded is proportional to the increase in the unit's size.

We need not include the entire gamut of demand in our analysis, however. Just as, in price analysis, profit-maximizing or money-

[3] Even changes in the size of the unit involve more than a simple quantitative change if the product is packaged. Imagine, for instance, tomato juice being made available to housewives in cans of various sizes ranging between one ounce and one ton, all selling at the same price (resale being prohibited). Surely the "best buy" would be some size smaller than the biggest.

revenue-maximizing firms would never be concerned with prices in that lower range in which the elasticity of demand is unity or less than unity, so, in vertical quality analysis, we can safely ignore the corresponding upper range of vertical quality, since cost-raising quality improvement would not be advantageous to firms in a fixed-price market unless it induced an increase in the number of units sold. In what follows, therefore, we need be concerned only with that range of outputs in which a rise in quality results in a rise in sales, and vice versa. We shall assume that

21. *Output demanded is an increasing function of the height of a product's quality, and is the same function for all products regardless of horizontal differences in quality.*

If all firms should raise (or lower) quality simultaneously, there would be no reason to expect any firm to gain (or lose) more or less than its proportionate share of the total increase (or decrease) in market demand. On the other hand, if one firm should make its product vertically superior to rival products, it could be expected to attract buyers previously attached to those rival products—and vice versa. We may reasonably assume, therefore, that

22. *When firms alter vertical quality in unison, their shares of the market remain unchanged; when one firm alters quality while others do not, its share of the market changes in the same direction.*

This raises the question as to whether vertical quality differences would be allowed to persist in a one-price market. Though this is obviously a question of considerable theoretical interest, the temptation to explore it will be resisted. It will simply be assumed that

23. *Every change in vertical quality made by a firm will be matched, either immediately or eventually, by all other firms.* (This could be true either because whenever one firm is prompted to alter quality vertically, others are similarly prompted for the same reason, or alternatively, because other firms feel forced to follow suit in order to retain their shares of the market.)

This assumption further narrows our field of inquiry. It may be defended on the ground that it is in keeping with the conclusion generally reached in monopolistic competition theory that when sellers of differentiated products have identical costs, every variety must sell at the same price in equilibrium.

Two other assumptions are made at the start, then later modified:

24. *Every change in vertical quality is matched immediately by all other firms, and each firm acts on the expectation that this will be so.* In other words, we are assuming here that the already-assumed "eventual" elimination of vertical differences occurs so quickly, and that this fact is so well recognized, that no firm considers it worth while to seek the temporary benefits of a newly created differential.

25. *The number of firms is fixed at ten.*[4]

Let costs—which are similar for every firm—be as shown in Figure 3. Each of the curves labeled *AC* depicts the average cost of producing a certain level of quality, designated by the subscript.

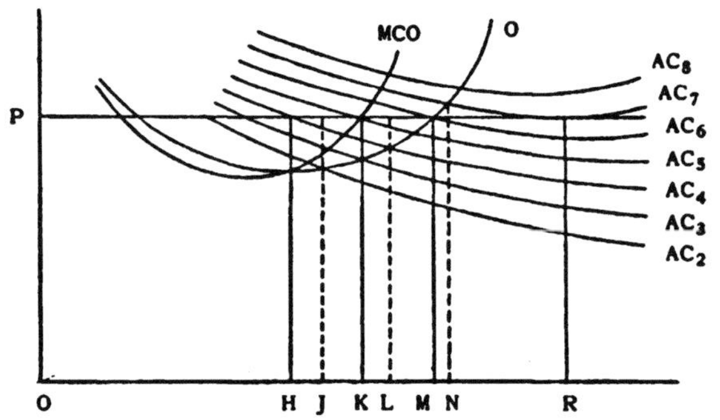

FIGURE 3

Since vertical quality is continuously variable, the family of cost curves drawn represents only a few of the possible heights of quality.

Let *OP* represent the fixed price, and AC_6 the cost curve applicable to the vertical quality currently being offered. Let the market demand for this quality at this price be such that every firm's demanded output (one tenth of the market demand) is *OM*. We thus start our analysis with each firm earning only normal profits. Our problem is: what variation will occur?

In order to investigate this question, we need further information about demand: namely, the change in quantity demanded associated with each change in the level of quality. This information

[4] Use of numbers rather than algebraic symbols to designate the numerousness of sellers in the market is solely for convenience in exposition, and is not to be regarded as diminishing the model's generality. If the reader prefers, n can be substituted for 10, $n + 1$ for 11, etc.

can be assembled in the form of a schedule, and plotted as a curve; but it will not be the familiar kind of demand curve showing quantity demanded as a function of price. Instead, it will show quantity demanded as a function of quality height, or, more precisely, of the cost uniquely associated with that height of quality at that rate of output.

Let us imagine the vertical quality to be varied while price is left unchanged, and note with respect to each quality height the output demanded, and also the unit cost of supplying that output. For each quality selected a point on our diagram can be found, the x distance of which represents the quantity demanded (the number of firms being what it is) and the y distance the unit cost of supplying that quantity. The locus of points so obtained is the desired curve.

This curve is sufficiently unconventional to warrant some further comments as to its character. It is, in a sense, a demand curve; for it shows, for each height of quality selected (represented by the cost of supplying that quality), the quantity demanded. But it differs from the ordinary demand curve in that it does not give the demand *price* but rather the demand *cost* at each rate of output— that is, the cost which the firm must incur in order to induce the purchase of that output, price being held at a predetermined level.

Its usefulness lies in the fact that it reveals the various options actually open to producers in a fixed-price market. In a "purely competitive" market the conventional average cost curve indicates genuine options. There each producer is at liberty to market as large or as small an output as he pleases; the cost curve and price line are useful in that together they reveal the consequences of each possible decision, and so show which options are less desirable than others. But in a fixed-price market in which products differ horizontally in quality the conventional average cost curve does not contain options actually open to producers. All but one of the cost-output combinations on the curve are unattainable as long as the buyers' demand functions, the price, and the number of firms remain as they are. The producer is not free to unload on the market any output he chooses of a given level of quality; only if he varies vertical quality to the extent necessary to induce buyers to purchase the output he has in mind does that output become a genuine option. And then it is the cost of producing *that quality* which is the relevant cost in weighing the desirability of adjusting output to that rate. Hence the curve just described, relating demanded output to the cost of supplying that output, serves precisely the same purpose in a fixed-price market which the conventional cost curve serves in a fixed-quality market.

Since it reveals the cost of each of the actual options open to producers, this curve will be termed the *cost of options* curve, or, more briefly, the *options* curve.

Referring back to Figure 3, suppose that if quality AC_2 is selected, each firm's share of the resulting market will be *OJ;* if AC_3 is selected, demanded output will be *OK;* if AC_4 is selected, *OL;* and if AC_7 is selected, *ON*. The curve *O,* drawn in conformity with these suppositions, shows the possible choices open to the firm.

What shape should an options curve have? If average cost curves were horizontal lines, the options curve would climb continuously upward, showing an increase in unit cost associated with each rise in quality. But since average cost curves are assumed to be U-shaped, a complication arises: vertical variation influences cost in two different ways. On the one hand it alters the cost associated with any given output—that is, it shifts the average cost curve bodily. Its effect along this channel is always the same: a rise in quality means a rise in the cost function, and vice versa. But vertical quality change also alters the actual rate of output (except when the elasticity of demand is unity), so that the unit cost becomes different from what it would have been if output had not been affected. In other words, it causes a movement along the curve. And its effect along this channel is not always the same. Up to a point a rise in quality brings an increase in sales that makes possible economies resulting in a lower unit cost compared with the cost of the same quality produced at the lesser rate; but this effect gradually diminishes in force as output is increased until there finally comes a point where diseconomies arise that raise the unit cost of producing a given quality. Thus for a while the two influences pull against each other; and it is likely that at very low rates of output the added cost of producing a higher quality will be more than offset by the reduced cost of producing on a larger scale. At such rates unit cost falls as quality is raised, and the options curve has a negative slope. But eventually the two influences reinforce each other, and unit cost must certainly change in the same direction as quality; the options curve then becomes positively sloped, and continues so until that height of quality is reached at which quality raising no longer raises output, when the curve becomes vertical. At still higher quality levels it bends backward.

If a family of U-shaped cost curves were drawn one directly above the other, it would be geometrically possible to draw an options curve which bent more and more in a counterclockwise direction until it showed unit cost falling once more as quality rose to very high levels. But this would be an improbable situation. It should be expected that when quality rises the average cost curve not only rises but shrinks toward the left. In the limiting case in which changes in vertical quality are merely changes in an accounting unit, the minimum-cost output rate must vary inversely with the size of the unit. In that case the locus of minimum points of the family of average cost curves would be a rectangular hyperbola, and no decline in output short of an actual decline in physical volume would keep the options curve from rising.

Since quality reduction is not advantageous unless it reduces unit cost, and quality raising is not advantageous unless it increases demand, we can confine our analysis to that range of vertical variation in which both unit cost and output vary positively with quality, thus ignoring the negatively sloped portions of the options curve.

It must be borne in mind that the position and shape of the options curve vary with the number of firms in the industry. This is because the firm's demanded output at each quality level varies with the number of firms, whereas its average cost curve relevant to each quality remains fixed. Therefore the firm's options curve shrinks upward and toward the left as new firms enter the industry. This complication will be dealt with shortly, when the number of firms is permitted to vary.

Now let us apply the options curve concept to the problem in hand. Under the assumed demand conditions it would not pay the firm to raise quality. At any output larger than *OM,* cost is higher than price. Quality reduction would be profitable, however. Any one of the four levels AC_2, AC_3, AC_4, and AC_5 would be an improvement. What level will be most profitable? The answer is found by drawing a *marginal cost of options* curve, showing for each rate of output the increment in total cost necessitated by a rise in quality sufficient to expand salable output by one unit. The point at which the marginal-cost-of-options curve intersects the price line denotes the output (and the quality associated with it) that maximizes profit. This is shown in Figure 3, the marginal curve being labeled *MCO*. It can be seen that every firm will select *OK* of quality AC_3. With a change to this new level supernormal profits are earned.

We now relax assumption 25, and allow the number of firms to vary. Profits invite entry. When an eleventh firm enters, each firm's demand shrinks to one eleventh of the market demand.[5] Its output, let us say, is now in the neighborhood of *OJ*; the firm is in disequilibrium once more. The options curve drawn in Figure 3 no longer holds. In order to determine the equilibrating adjustment needed, a new curve must be plotted, appropriate to an eleven-firm industry. In fact, we need at this point to visualize a whole family of options curves, one for each of the possible number of firms.

How should such a family be drawn? If we think of the options curve as composed of points, each of which lies on some *AC* curve associated with some level of quality, a new firm's entry causes these points to slide clockwise along their respective *AC* curves sufficiently so that each point is the same fraction of its former distance from the vertical axis. For each number of sellers these points lie differently and a separate options curve can be drawn.

None of the curves will intersect another; for when the number of firms increases, each point of the newly formed curve must lie somewhere to the left of the corresponding point (that is, the point representing the same height of quality) on the old one. And since the *AC* curves are concave from above, and descend more steeply than the options curve, and rise less steeply, each point on the

[5] It does not necessarily shrink to ten elevenths of its former demand. A somewhat smaller shrinkage is to be expected, since there is an offsetting factor. Entry of a new firm causes products to be "spaced closer together" horizontally, which may be presumed to reduce their horizontal unsatisfactoriness on the average, and thus to increase the market demand. But so long as market demand grows less than sufficiently to offset completely the shrinkage in the firm's percentage of the market, the firm's demand will become smaller. This must now be assumed to be the case.

options curve near its right extremity will move downward and to the left when a new firm enters, but the slope of its path will be less steep than the slope of the options curve at that point, while each point near the left extremity will move *upward* and to the left, and the slope of its path will be steeper than the slope of the options curve at that point. Therefore any new options curve formed by the entry of an additional firm must lie wholly within the area bounded by the former options curve.

The curves will also converge as we follow them upward to the left, since they cannot cross the y axis but must continually approach it. And they will also converge as we follow them upward to the right, since each curve in the family must approach the limiting output at which the product must be given away free (or its size made infinitely great if the price remains fixed) and this limit must also be the y axis, that is, an infinitely small number of units.

The family of options curves appropriate to our model is shown in Figure 4. In this Diagram O_{10} corresponds to the O of Figure 3, and MCO_{10} to MCO, the subscript 10 indicating a ten-firm industry.

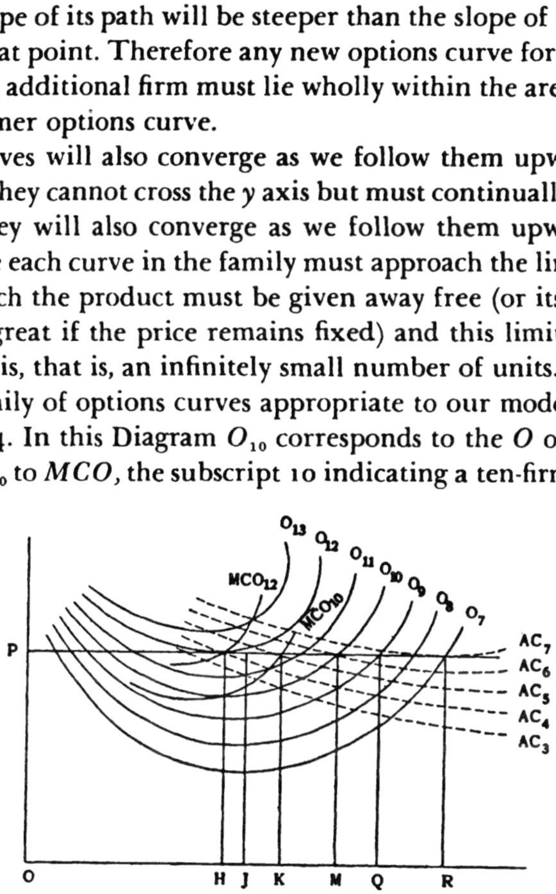

FIGURE 4

The new output rate is shown in Figure 4 by the intersection of AC_3 and the new options curve O_{11}. It will be profitable for each firm to move to the right along O_{11} (that is, raise quality) until it reaches an output at which its new marginal-cost-of-options curve (not shown) intersects the price line. As O_{11} has been drawn, excess profits will continue to be earned. These will attract further entry, which will cause a further shrinkage in output per firm, which will induce firms to adjust quality upward still more; and this process must continue until excess profits are eliminated. Curve O_{12} is drawn to illustrate the final equilibrium position. With twelve firms in the industry OH is the equilibrium rate of output, and AC_4 the equilibrium height of quality. Thus we see that long-run equilibrium is reached when the options curve shifts upward and to the left (as new firms enter) until it becomes tangent to the price line.

This is the maximum number of firms (and varieties) possible in the industry. If a thirteenth firm entered, there would be no possible way by which firms could avoid losses.

6. ALTERNATIVE EQUILIBRIUM SITUATIONS

Different equilibrium positions are reached if assumptions 16 and 24 are altered. Suppose we assume that each firm expects the vertical quality of all rival varieties to remain unaltered (although actually, since all firms are faced with similar conditions and behave similarly, all firms make identical decisions as to vertical quality). Now the firm's actions are based on the belief that any given cheapening of quality results in a greater loss of sales than before, and any given quality betterment in a greater gain in sales. The options curve concept can be retained, but new curves would have to be drawn, indicating, for each industry-wide level of quality and for each number of sellers, the sales volume which the firm would secure by each vertical change if other firms kept quality unaltered.

This is a more complex situation. For each number of sellers in the industry, a family of options curves is needed, each of which intersects the "industry-wide options curve" (the curve shown in Figure 3) at the point at which the quality of the firm's product is the same as that of the rest of the industry. Let us imagine, in the situation depicted in Figure 3, that one firm varies quality while all other firms adhere to quality AC_3. How should the options curve appropriate to this supposition be drawn? It would intersect the price line where the AC_6 curve intersects it, but it would be a flatter curve—it would intersect the AC_5 curve, for example, further to the left, and the AC_7 curve further to the right—and would be tilted more to the right, so that its minimum point would lie further to the right.

Provided that the newly drawn curve were not so tilted as to be tangent to the price line, it would show the profitability of the firm's reducing quality to some lesser extent—to AC_5, say. Since all other firms likewise reduce quality to AC_5, the firm finds that its output has not shrunk as much as had been anticipated; their action has shifted the firm's options curve downward and toward the right so that it intersects the AC_5 curve at the point where the industry-wide options curve in Figure 3 intersects it. Further quality reduction is now worth while. By a succession of such adjustments the industry reaches its short-run equilibrium, with some excess profits being earned, and entry of new firms invited.

The process of reaching long-run equilibrium involves similar adjustments. These need not be described here in detail. It is sufficient to point out that because of the tiltedness of the options curves (as compared to the industry-wide curves shown in Figure 4), tangency to the price line will be achieved before the number of firms is increased to twelve; in the final equilibrium quality will be higher than AC_4 and the number of firms fewer than twelve.

Now let us replace the profit-maximizing assumption (16) with the assumption that firms desire no more (and no less) than an average return on investment, but always prefer a larger to a smaller volume of sales. Their objective is thus output maximization, subject to the condition that average cost does not exceed price. In this case, contraction of output through quality reduction for the sake of temporary profit will be avoided. Expansion of output through quality raising is also undesirable so long as the number of firms remains unchanged, since it entails losses. But if we suppose that each firm believes that a rise in quality will force some competitor out of the market, and is willing to pay a temporary price for eventual gains in sales, quality raising will occur.

If firms, producing OM of AC_6 in a ten-firm industry (see Figure 4) raise quality to the level at which their new cost curve intersects the price line at, say, an output of OQ, and if the market demand should be nine times this output when quality is at that height, sales would temporarily fall short of *OQ* and losses would be incurred until one firm left the industry; then the remaining nine would earn the "normal" rate of return once more, and on an increased volume of sales. This process would continue either until the minimum point on an average cost curve were reached, or, if that point could not be reached in an industry of more than one firm, until the number of firms were reduced to one. According to Figure 4, a seven-firm industry could produce quality AC_7 at the minimum average cost rate, *OR*. Here equilibrium is reached. This is the maximum height of quality that can be produced and sold at price *OP* without incurring losses.

If firms sought to maximize output, yet were unwilling to undergo temporary losses for the sake of growth, no firm would follow this aggressive policy, but would adjust quality to the height at which the unit cost of producing the quality demanded equaled the price. Thus *OM* would be one of several possible long-run equilibrium positions. Which position was actually reached would depend on the number of firms that happened to be in the industry.

By varying our assumptions we have found several alternative sets of conditions of long-run equilibrium: (1) that in which the number of varieties (among those numbers capable of being produced at a cost equal to price) is maximized, quality height being adjusted to the level necessary to achieve this maximization; (2) that in which quality height (among those heights capable of being produced at a cost equal to price) is maximized, the number of varieties being whatever number makes possible this maximization; and (3) various intermediate positions, in which vertical quality is higher than in the first case though less than in the second, while the number of varieties is greater than in the second case though less than in the first. We have located a *range* of equilibrium positions.

That there is a range rather than a single position is not due to the fact that competitive adjustments are forced to take the form of quality changes rather than price changes. The equivalent of each of these positions can be shown on a conventional diagram of a firm selling a differentiated product of fixed vertical quality in a monopolistically competitive market.

Equivalent to the case of concurrent quality changes is the case of concurrent pricing, which, when employed by profit-maximizing firms, maximizes joint profits. Each firm adjusts its price on the basis of a demand curve which is 1/*n*th the industry demand curve, *n* being the number of firms. In Figure 5 such a demand curve is shown as D_{10}. If firms have been producing for a market in which

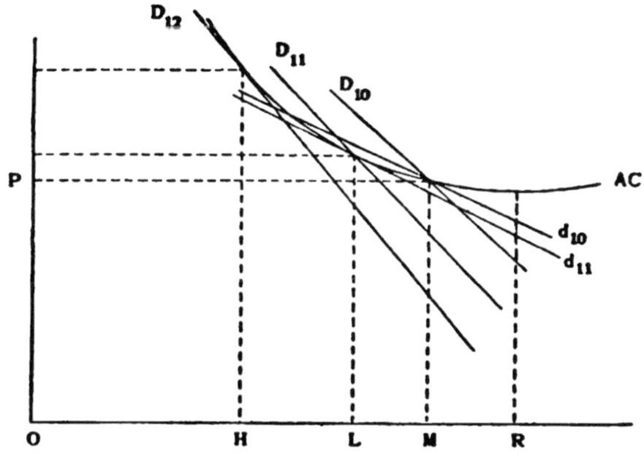

FIGURE 5

price has been fixed at *OP* and are now free to alter the price, they can gain supernormal profits by raising the price. But entry of new firms will shift the demand curve leftward. Equilibrium is attained at *OH* when the curve reaches the position D_{12} where it is tangent to the cost curve.

The equivalent of our second equilibrium position can also be shown. Its conditions correspond to the Chamberlinian assumptions. Each firm's demand curve (based on the expectation that prices of other varieties remain unchanged) is more elastic than the curve showing the volume of sales when all competitors' prices change too. Equilibrium is attained at *OL* when entry of new productive capacity pushes D_{10} leftward to the position D_{11}, where the firm's demand curve, which was d_{10} and is now d_{11}, becomes tangent to the cost curve.

The equivalent of our third equilibrium position, based on desire to maximize output, acceptance of average-cost pricing as a norm, and willingness to take temporary losses, is at *OR*.

These comparisons enable us to see how similar to price variability in its final effects is vertical quality variability.

The plurality of these equilibrium positions raises an interesting question. Which of them is best from the standpoint of the community's welfare? Which provides the highest aggregate amount of buyer satisfaction?

Conventional theory provides a ready answer. Only when firms are producing at the least cost point, and when price is made equal to this cost (or, in terms of the kind of market being examined here, when quality is raised until the least cost point coincides with the fixed price), are buyers getting the most for their money. At any of the equilibrium positions to the left of *OR*, the unit cost of producing the quality associated with that position is higher than the minimum for that quality. Alternatively, the quality is lower than it would be if firms were forced to produce at the "optimum" rate and to raise quality until excess profits vanished. It is usually argued that these equilibrium positions to the left of *OR*, which are of precisely the sort monopolistic competition theory has taught us to expect in markets for differentiated products, leave buyers "worse off" than they might be.

Yet the answer is not so simple as that. Buyers are worse off in one respect, since quality is vertically lower than the highest consistent with long-run equilibrium. But if the average satisfactoriness of the products bought varies with the number of varieties offered, buyers are better off in another respect, since with more varieties to choose from, wants are satisfied on the average with greater precision. Even the extreme position at *OH* would be an improvement over *OM* if the added satisfaction gained by consumers as a result of there being twelve varieties to choose from instead of ten happened to outweigh the decrease in satisfaction resulting from the cheapening of quality.

Have we any way of knowing whether the gain is as great as or greater than the loss? No precise answer seems possible, but a few things can be said that will throw some light on this question.

If buyers were completely indifferent to horizontal quality variation, deriving just as much satisfaction from one variety as another, but preferred higher to lower quality, the minimum point on the highest average cost curve that touches the price line (output *OR* of quality AC_7) would clearly maximize their satisfactions. But under these assumed conditions no other position could be one of equilibrium. If firms were producing a quality inferior to AC_7, and if one firm raised the quality of its product by some slight amount, it would attract buyers from every corner of the market; such a firm could sell all it wished to, and therefore would select that output which maximized its profit. Any other firm that followed suit could similarly take customers away from all firms that failed to follow suit. Competitive raising of quality would be forced on producers, and would continue until the quality level was reached at which the average cost curve was tangent to the price line.

At the other extreme, if buyers were completely indifferent to vertical changes above the quality level AC_5, and if their satisfaction with any product varied with the closeness of its horizontal quality to their respective optima, the product-maximizing position (output *OH* of quality AC_5) would clearly be best, and *OR* could not possibly be an optimum position. And here too it is easily shown that no other position could be one of equilibrium under the conditions assumed. If buyers were indifferent to vertical quality differences in the range above AC_5, cost-raising quality improvement would gain nothing, and firms would not resort to it. So long as the industry contained fewer than twelve firms, each firm would earn supernormal profits. This would attract entry. Only when the number of firms (and varieties) had risen to twelve would further entry cease.

No intermediate position (between *OH* of twelve varieties of quality AC_5 and *OR* of seven varieties of quality AC_7) could be either an optimum position or an equilibrium position unless consumers considered *both* horizontal and vertical differences significant. If such were the case, the optimum position would depend on the relative strengths of the two elements of preferability. *A priori* it seems plausible to suppose that there is one position which maximizes buyers' satisfactions. This would be true if equal increments of unit cost when applied either to quality-raising or to increasing the number of varieties brought diminishing increments of aggregate satisfaction. In cases where quality-raising is actually a form of quantity-raising, it seems logical to expect diminishing psychic returns. Even in cases where the changes are fundamentally qualitative, as in cost-raising improvements that make a watch a more accurate timekeeper, there must come a point where further improvement becomes less and less important. Likewise, if the number of varieties is increased indefinitely, there must come a point where each new addition increases consumers' aggregate satisfaction by a smaller amount than did the previous addition. And eventually, as firm after firm enters the field, each firm's scale of production will shrink to the point where each further shrinkage raises unit cost by increasing amounts. If both kinds of improvement yield diminishing psychic returns, there must be some position at which the increment and decrement of satisfaction produced by a double adjustment are equal. This will be the position of highest aggregate satisfaction.

But this statement must be qualified: it is meaningful only to the extent that interpersonal comparisons are meaningful. Any change in the number of varieties accompanied by a compensating change in quality height (or price) must improve the welfare positions of some consumers at the expense of others. If vertical quality has been maximized, those consumers whose wants are being satisfied with perfect precision by the few varieties being produced are as well off as they can be. If now an additional variety is introduced, and if the locations of existing varieties along the horizontal quality spectrum are altered in order to make room for the new variety, and if vertical quality is lowered (or price raised) simultaneously, these particular consumers are injured while others are benefited.

7. IMPLICATIONS OF THE ANALYSIS

The writer is fully aware of the limited scope of the analysis just presented. The model studied illuminates only one small corner of the great arena of quality competition.

By restricting ourselves to a study of *equilibrium*, we have obtained a picture of competitive behavior with regard to quality variability which is highly artificial and seriously incomplete. This is unavoidable. A prerequisite of equilibrium is the attainment of perfect knowledge (or, alternatively, an unalterable state of knowledge). Equilibrium analysis therefore turns its back on the semi-ignorance, uncertainty, and exploratory character of real-life economic activity. The typical patterns of quality competition are in large part founded on disagreement as to the best means of satisfying wants, and reflect the continual experimentation that is so strikingly characteristic of economic behavior. Innovation is intelligible only when viewed in such a setting; it is inherently experimental, risky, a challenge to the current state of knowledge and currently prevailing attitudes. Equilibrium analysis, on the other hand, necessarily assumes that the newest ideas have become common property, and have been tested out long enough so that their value is no longer in doubt. It therefore omits this important aspect of economic behavior, revealing only those aspects of quality competition that are connected with cost differences and the diversity of wants.

Our analysis would also have been more comprehensive if it had not abstracted from certain frequently found conditions in the real world—for example, semi-ignorance, gullibility, and the bewildering complexity of choice when quality variability is multidimensional and information about quality and product performance is not readily available.

Nevertheless, the analysis in this chapter seems useful on two counts. First, by focusing our attention exclusively on the quality dimension of competition, it enables us to obtain a clear view of a type of maximizing, equilibrating behavior that is obscured by the conventional techniques of economic analysis. Even though this behavior occurs typically in conjunction with price-output behavior, it is a useful analytic device to study it in isolation. Second, the prevalance of unchanging prices in many areas of the American economy today makes an analysis of market behavior under conditions of price inflexibility especially pertinent to contemporary economic problems.

Our analysis has shown that horizontal quality competition performs certain functions which price competition cannot perform. It causes producers to select those qualities in a product that satisfy wants more precisely—that is, it tends to minimize the unsuitability of products to the uses to which they are put. It broadens the range of choice open to buyers, and provides the variety that people crave.

Our analysis has also made clear one thing which may be familiar enough to observant students of economic life but which customary methods of economic analysis do not show: namely, that vertical quality competition performs essentially the same functions that price competition performs. Whenever price is excessive relative to product—or, to put it the other way round, whenever product is deficient relative to price—it causes producers to revise their offers, making them more attractive. Thus, like price competition, it induces behavior which results in the elimination of excessive profits. Similarly, the firm's equilibrium position in a fixed-price market with vertical quality freely variable appears to be identical with the position reached in a fixed-quality market with price freely variable. When products are horizontally differentiated, and when firms seek to maximize profits, the firm's output falls short of that at which unit cost is minimized. When horizontal differentiation is lacking, output settles at the minimum-cost rate. Thus vertical quality competition, like an able understudy, steps into the breach when price competition ceases to function and brings about results usually associated only with price flexibility.

Our venture into Looking-Glass economics has thus made it evident that price theory provides an incomplete analysis of market situations. The extent to which the results associated with competition are realized in any market cannot be correctly assessed without taking into consideration conditions with respect to quality differentiation and variation.

Chapter XII

QUALITY AND PRICE COMPETITION COMBINED

IN the two preceding chapters quality competition was examined in isolation. It is not necessary for us to make a similar examination of price competition in isolation, since modern price theory has already accomplished this task. What is now required is a synthesis: an examination of what happens in markets in which both price and quality are free to vary.

1. MARKET BEHAVIOR UNDER COMPLETE COMPETITION

Three kinds of quality variation and differences have been distinguished: horizontal, vertical, and innovational. In markets in which price as well as quality is free to vary, producers who wish to maximize their positions may find it advantageous, at one time or another, to vary quality in any or all of these ways. In addition, producers can take two other kinds of action.

It was shown in the last chapter that when price is inflexible, and when price is out of line with cost, the situation can and will tend to be corrected by the making of cost-raising or cost-reducing vertical quality adjustments. In markets in which both price and quality are variable, this method is still available, but producers now have an alternative, more direct method: price change. Producers are free to employ either method separately, or both simultaneously.

It will not be a matter of indifference which method is employed, except in the very rare case in which different quality levels or package sizes are equally convenient. Typically it makes a good deal of difference to buyers how "much" is embodied in each unit. If the product is a durable good, vertical variation amounts largely to variation in the degree of convenience or luxuriousness with which the product renders its service, and people are likely to have definite ideas as to how much of their incomes they wish to spend for the particular service or activity for which a product is used. If the product is of the kind that is consumed in use, vertical variation may be of the sort that enables a changed amount to render an unchanged service (for example, superior coffee that makes more cups per pound); in this case the ideal quality level is whatever is most economical. Or it may consist of variation in the size of the package. Generally, the larger the amount contained in the package, the cheaper the contents per unit of weight or volume; but larger containers are less convenient to store and may result in waste or spoilage of seldom-used goods. Again buyers are likely to have definite preferences, each finding some size best suited to his needs. An advantage of permitting both price and vertical quality to vary is that only through their simultaneous variation can buyers indicate their preferences for the most suitable package sizes and the most desired level or levels of luxuriousness of durable goods.

Completely competitive markets have another characteristic not found in single-fixed-price markets: persistent vertical quality differences, accompanied by corresponding price differences. Two reasons can be advanced to explain such differences. First, people's incomes are dissimilar; therefore some buyers can afford to spend more on a given activity than can others. And second, even among people of equal incomes, some persons devote larger amounts of time, energy, and interest to a given activity than others, and will consequently find it advantageous to devote a larger fraction of their income to the activity in question than would a person who engaged in the activity only very infrequently.

Persistent price differences may arise occasionally for another reason. At the fringes of the horizontal quality scale may lie qualities which appeal to only a few buyers; yet the appeal may be so strong that these buyers are willing to pay a higher price, as compared with other varieties of equivalent grade, in order to obtain the favored variety. If the higher price just covers the higher unit cost of small-scale production, there will be no tendency for the difference to disappear.

In brief, the two new kinds of producer action are: (1) the making of price adjustments within each quality level, associated with cost-reducing innovations in production methods, or with changes in the prices of inputs, or with variation in the intensity of competition, and (2) the establishment of persistent price differences associated either with the existence of different quality levels or package sizes or with products of unusual quality in small demand.

The fact of price differences introduces a new element into consumer choice. Consumers will of course seek to maximize their satisfaction, as they did in single-fixed-price markets. Their procedure, however, will be somewhat more complicated. In a single-price market each consumer's criterion of choice is very simple: he merely asks himself, as it were, "Which of these competing products comes closest to 'hitting the bull's-eye' of my basic wants?" But in a multi-price market differences in quality must be weighed against differences in price. One method of analyzing this weighing process is to think of the consumer as taking the higher-priced product of any pair as his standard of comparison, and asking himself "Would I prefer to have Product A, or to have Product B and, in addition, C dollars of income to spend on other things?" [1] In a multi-price market, therefore, it may happen frequently that the product which best satisfies a person's basic wants, price considerations aside, is rejected in favor of some less expensive nonoptimal product.

Cost changes, quality changes, and price (or output) changes in completely competitive markets may be described as being impelled by the desire to achieve a maximum position, and as being in this sense equilibrating movements. But such movements frequently are also disequilibrating. We have already seen that a horizontal quality adjustment made by one firm tends to put neighboring firms in the industry into disequilibrium, which can be corrected only by quality adjustments on their part. The entry of a new firm offering

[1] Cf. Ruby T. Norris, *The Theory of Consumer's Demand*, pp. 114–29; John Bates Clark, *The Distribution of Wealth*, pp. 235–41.

a new variety is similarly disequilibrating. A new idea, previously unknown or untried (and therefore unaccepted) by the industry in general, produces a similarly dual effect: it is in one sense equilibrating, since it puts the firm which adopts it into a better position; but it is also disequilibrating, in that other firms are now in a worse position than before. Thus an innovation, whether in a firm's production method or in its product, lifts slightly the veil of ignorance which keeps costs higher and products inferior to what they would be in the millennium, and generates a movement toward that very-long-period equilibrium; but it also creates an intolerable differential in cost or quality, and thus necessitates a whole series of lagged readjustments on the part of competing firms. This ceaseless series of changes which are both equilibrating and disequilibrating is mentioned at this point because it is such an important feature of real-world exchange, yet one which disappears from the economist's sight as soon as he turns his investigation to the problem of the conditions of equilibrium.

2. EQUILIBRIUM UNDER COMPLETE COMPETITION

Only a very brief examination of the conditions of complete equilibrium will be attempted here. It will be assumed, as in the last chapter, that the kind of market under examination is one in which the product can be varied both horizontally and vertically, and in which buyers' basic wants are so various that the number of significantly different varieties wanted by them exceeds the number that could be produced by firms operating at minimum average cost. An unchanging technology is also assumed. All possible innovational changes within this given technology have already taken place, and sufficient time has elapsed to wipe out all the temporary differential advantages in cost and quality that accrued to the enterprising producers who were ahead of the general procession. Any differentials now remaining must be regarded as due to differences in human ability or geographical location or fertility of resources, and will be reflected in rents or differential wages.

Every business firm will seek to establish itself at the "quality location" offering the maximum advantages. What a firm considers most advantageous depends upon the particular goal it seeks—whether, for example, its goal is maximum net revenue, or maximum sales volume, or maximum shelter from close substitutes (that is, minimum elasticity of demand), or some combination of the three, or some other goal or goals. However the advantages are defined, if all competing firms seek the same ones the result will be a tendency to equalization of the advantages associated with each location. If profits are sought, disparities in the profit rate will tend to be eliminated; if sales volume is the objective, no producer can expect to enjoy lastingly a quality location far enough removed from substitutes to permit a markedly above-average sales volume; if all producers seek to maximize the "degree of monopolistic exploitation" as measured by the smallness of their elasticity of demand, those producers who face highly elastic demand curves will tend to edge away from their overly "competitive" locations and toward those locations which are temporarily enjoying the greatest seclusion. In each case the result is a movement toward equalization of the advantages sought.

If producers were forced to cluster so as to increase the degree of price competition—this could be accomplished by establishing, either by law or by trade association edict or by custom, a limited number of "standard" qualities and prohibiting the sale of "nonstandard" products—the effect would be to increase the elasticity of each seller's demand curve, and thus make the market resemble more closely a complex of "purely competitive" markets. Yet, looking at the matter in another light, the effect would be to perpetuate disequilibrium. Vacancies in the chain of substitutes would be created. Except for the institutional barriers erected, entrepreneurs would have golden opportunities to move into the unpopulated "locations" and earn excess profits by catering to those wants which now remain neglected. And for those buyers whose preferences lay in the neglected areas, the denial of an opportunity to buy a product whose quality was for them more nearly optimal than any now being produced would represent a real loss of potential welfare.

This means that in a world such as ours a system of "purely competitive" markets cannot be in equilibrium. Pure price competition in every market is attainable only at the expense of limiting the total number of products. To do so would create gaps in the chain of substitutes. The demand curves for the unproduced varieties would have negative slopes, and would thus offer opportunities for "monopoly profits." If producers were perfectly free to vary their products as they saw fit, it would be to the advantage of some producers and some consumers to shift some of the economy's resources into the production of those unproduced varieties. So long as this flow of resources is advantageous, yet has not taken place, the economic system cannot be in equilibrium. True equilibrium is that state of affairs in which no firm can move to a new position, *either already occupied or unoccupied,* more advantageous than its present one.

What can be said about the equilibrium number of varieties in any given industry? This question was examined in the previous chapter; but now it can be restudied under more propitious circumstances. Instead of having to make the cumbersome assumptions of price rigidity and a backhanded quasi-price-flexibility achieved by vertical quality variability, we can attack the problem in a more direct and natural manner. We shall assume price variability, decreasing costs throughout the practicable range of production scales, horizontal quality variability in one dimension, and consumer attachments reflecting the diversity of basic wants. We shall also assume that every producer follows a policy of average cost pricing—or, more precisely, if we take into account the fact that demand responses to price or quality changes are not instantaneous, that every producer selects the price which will be equal to average cost for the volume of sales which that price and quality will eventually call forth. This will not lead to conclusions very different from those reached by assuming profit maximization, yet it will simplify matters, and will focus our attention on the important point: that under conditions of decreasing costs a larger volume is associated with a lower price, and vice versa.

These assumptions provide a framework within which various mathematical models can be constructed based on various assumed relationships between the variables concerned. No such models will be constructed here, but their general nature can be described.

The following functional relationships would have to be given quantitative content:

1. For each consumer, or class of consumers of given preferences, the quantity demanded of each product is some function of its price,

2. and also of its distance from that location on the quality spectrum which, in the opinion of those consumers, is optimum.

3. Each consumer is assumed to have his own valuation or indifference system by which he can determine, with respect to any pair of "neighboring" products, the difference in price between them that would just offset the difference (to him) in quality. This money equivalent is an indication of the strength of his attachment to the preferred product. Its value must be assumed to vary with the distance between the two products. But rather than assume that all members of each consumer group have identical preference systems (which would make the system highly unstable, and would also be unnecessarily unrealistic), it seems better to assume a functional relationship between the amount of price difference between any two neighboring products and the fraction of the consumer group formerly loyal to the higher-priced variety which has transferred its allegiance to the less desirable but lower-priced variety. A workable function would be one in which equal increments of price difference gain progressively fewer new customers for the cheaper product.

4. For each variety produced, price (being equal to average cost) is a decreasing function of the output; and all producers have identical cost functions.

5. The density of the buying population at each location on the quality spectrum is a function of its distance from the center of the spectrum. Or, alternatively, the density can be assumed to be constant.

Even without constructing an actual model having these specifications, one can surmise that the two most important elements are (*a*) the strength of consumer attachments to particular varieties, and (*b*) the steepness of the descending slope of the average cost curve. If consumer attachments are weak and if curtailment of output leads to sizable increases in unit cost—and hence price—one can expect less proliferation of varieties than would be found if there were strong horizontal attachments and little price benefit resulting from larger-scale production.

A conclusion to be drawn at this point is that in industries in which quality differences and consumer attachments are important elements, the equilibrium number of firms in the industry is determined, not by the minimum-unit-cost scale of production for each firm, but rather by the relative strengths of the two forces just described. It is variety of preferences, not the fact that cost curves eventually rise, that keeps a number of producers in the field.

This conclusion is not intended to imply that the equilibrium number of varieties of any given product needs to be very large—simply that it may be large enough to prevent production at minimum average cost. It seems likely that a moderate number will usually be sufficient to satisfy with reasonable precision the diversity of tastes encountered in the real world. Price theorists are agreed that pure price competition requires "very large" numbers of competitors—with no top limit. Effective quality competition, on the other hand, does have a top limit, which is determined by the buyer's capacity to discriminate intelligently. And it may well be that numbers far below this limiting ceiling provide a range of choice that is tolerably satisfactory for all concerned.

This conclusion is not fundamentally different from that of standard monopolistic competition theory; the difference is one of emphasis and interpretation. The conventional diagram of the monopolistically competitive firm in long-run equilibrium, showing the demand curve tangent to the average cost curve, can be interpreted as showing a balancing of the same two forces. Modern price theorists usually regard the slope of the seller's demand curve as a measure of his monopoly power; but looking at it from the buyers' viewpoint, is not this monopoly power simply a reflection of —and hence a measure of—buyers' reluctance to desert a preferred quality in order to take advantage of a cheaper price? To the right of the point on the demand curve representing the current price, the extent to which the curve falls indicates the unsatisfactoriness of the product itself in the eyes of many consumers whose attachments lie elsewhere, while the extent to which the curve rises to the left indicates the unsatisfactoriness of nearby substitutes in the opinion of many of the product's present buyers.

What implications does our analysis have for a society in which the multiplicity of wants and preferences creates in all industries the type of situation just described? Suppose an economy composed of *n* industries, in each of which *m* different varieties of the product are wanted and no more than *m* are capable of being produced. Furthermore suppose complete symmetry: every variant of every

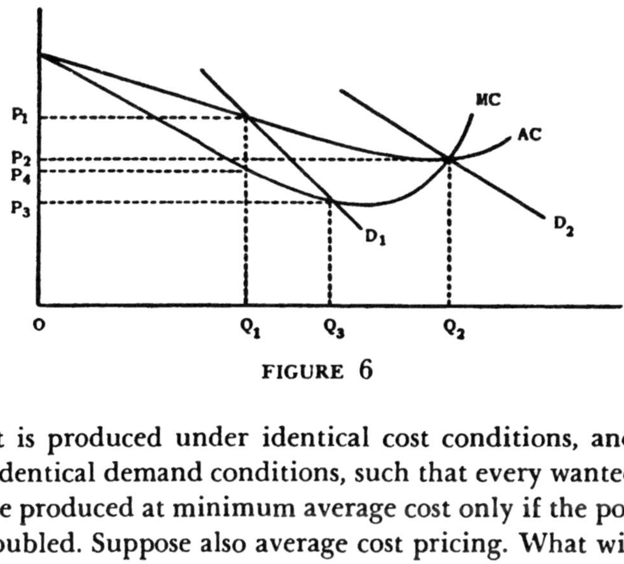

FIGURE 6

product is produced under identical cost conditions, and is sold under identical demand conditions, such that every wanted variety could be produced at minimum average cost only if the population were doubled. Suppose also average cost pricing. What will be the equilibrium number of products?

Only if preferences for one quality as opposed to another are sufficiently slight will the equilibrium number be no more than 1/2 *nm* and will each be produced at minimum average cost. And only if buyers' preferences are extremely strong will the equilibrium number be as high as *nm*, the higher real costs of production being reflected in higher prices. Otherwise some intermediate situation will be true. Let us say that buyers' attachments are sufficiently strong to make the number of varieties in each industry exceed 1/2 *m* by 50 per cent. This means that each firm is producing under conditions as shown in Figure 6. Its demand curve is D_1, its output is OQ_1, and its price OP_1. An output of OQ_2 at price OP_2 is impossible. It could become a reality only if attachments were sufficiently weakened to eliminate a sufficient number of varieties to augment the demand for each of the remaining varieties to D_2. (According to Figure 6, output of each variety can be doubled if the number of varieties is reduced by one third, since the amount of resources needed to produce each unit is reduced by one fourth.)

It is interesting to raise the question whether such a competitive equilibrium corresponds to the position of maximum welfare.

Would marginal cost pricing improve matters? Actually, no. Suppose universal marginal cost pricing, the government subsidizing each firm to the extent of its losses, and obtaining the necessary funds by taxing consumers in proportion to their expenditures. It might be thought that this would lead to an expansion of each firm's output to OQ_3 and a reduction of price to OP_3. Yet that could not occur. No firm's output could be expanded unless there were a contraction of output somewhere else. (Full employment is assumed.) The demand curve D_1 in Figure 6 does not show this, since it is an ordinary one, drawn on the assumption that other prices remain unchanged. But it is at once apparent that this curve is not relevant to the present problem. A fall in the prices of other goods would cause D_1 to shift to the left. If all prices were lowered in unison, no commodity would become cheaper relative to other commodities, and there would be no tendency for substitution to take place. With each change in price, therefore, D_1 must move in such a way as to show an unchanged quantity demanded. (In other words, a demand curve for the firm drawn on the assumption that all firms in the economy alter prices concurrently would have to be a vertical line.) Marginal cost pricing, then, would reduce the price to OP_4 and leave output unchanged at OQ_1.

It might be thought that the reduction in all prices would raise real incomes, creating a counter tendency for D_1 to shift to the right. True, income *before taxes* would be raised. But if the amount by which total revenues fell short of total costs were collected in taxes and paid as a subsidy to producers, the reduction in disposable incomes brought about by taxation would in the aggregate exactly negate the rise in real incomes which would otherwise have occurred. And since the added taxes are proportionate to consumers' expenditures, there would be no change in the allocation of resources and the number of commodities in the economy. The only thing new would be the tax-and-subsidy mechanism by which producers would be enabled to collect indirectly what they would not be permitted to collect directly from their customers.

This does not demolish the case for marginal cost pricing. If demand and cost functions differ markedly from industry to industry, the allocation of the economy's resources will be substantially different under marginal cost pricing from what it would be under average cost pricing. Yet the analysis just given does reveal an important qualification that needs to be attached to the welfare economists' proposition that a firm's output should be sold at the price at which the marginal cost and demand curves intersect. This proposition is valid only for economic systems in which most industries are purely competitive (or virtually so) and "natural monopolies" are exceptional. If the contrary is true—if the typical situation in an economy is production under constant or declining marginal cost—the demand curve for a typical seller's product is an incorrect indicator of the extent to which the so-called monopolists "restrict" output.

Would the enforced elimination of some varieties be in the public interest? In answering this question, it may be instructive to suppose a very simple model. Let us imagine a community in which there exists a "recreation" industry consisting of four firms, each producing one of the four possible kinds of recreation equipment: tennis, golf, fishing, and skiing. These four "varieties" of equipment are the only ones capable of being produced. The public's tastes are such that the population is evenly divided in its preferences for the four kinds of recreational activity; and each person wants to engage in one, and only one, type of sport. Due to the society's advanced technology, the cost functions of each firm are such that it produces at minimum unit cost only when its scale of production is large enough to serve half the population. Herein lies the dilemma: only one fourth of the population *wants* that kind of recreational equipment. The result is that there are "too many" firms in the industry, each producing at "too small" a scale and "too high" a price. As economists we are concerned with the problem of achieving maximum efficiency in the satisfaction of the community's wants. Can matters be improved? If skiing and fishing were outlawed, the devotees of these sports (we shall assume) would reluctantly turn to golf and tennis, and production at minimum cost and price of these two products would be assured. But the net result would be merely to increase the welfare of some members of the community at the expense of others. It cannot be claimed with validity that the community is made better off by outlawing any two of the sports.

Now let us suppose one hundred identical communities like the one just described, which have previously existed in isolation but are now given the opportunity to exchange with each other; and let us neglect the costs of transportation and suppose no tariff barriers. There are now four hundred firms in the recreation industry. Each firm's market has now increased one hundredfold. And each firm is no longer a monopolist of its own variety of recreation equipment, but faces ninety-nine competitors. It can afford to expand output. The result will be a wave of expansion and price-cutting; in the end each of the four kinds of sports equipment will be produced by fifty firms of optimum size, each operating under conditions of pure competition.

Why is the outcome different in the second case from what it was in the first? The difference is not due to the presence of oligopoly behavior in the isolated community; though a four-firm industry was supposed for the sake of simplicity, the oligopoly complications were avoided by the assumption of average-cost pricing; our first model therefore illustrates the case of product differentiation or monopolistic competition rather than oligopoly. Yet the difference between the two cases *is* due to fewness. The essential reason why the second outcome differs from the first is because *the population has been increased* whereas *the number of different products* wanted by members of the population *has remained unchanged.*

One moral of this tale, which has been made familiar to us by Adam Smith, is obvious: Large markets are needed in order to reap the full advantage of the economies of specialization and mass production. Another moral is less obvious, but perhaps equally important for us of the twentieth century: The proliferation of wants and activities in our complex society, the cultivation of enhanced powers of discrimination, the desire to satisfy wants with a high degree of precision, and technological advances in the exploitation of possibilities for specialization and mass production have combined to bring our civilization to the point where it can no longer expect to reap the full advantages of mass-production techniques except by the dubious expedient of curtailing people's wants and activities.

Proceeding with this line of thought, the difference between actual unit cost and minimum unit cost of the typical products of our economy is a measure of the price we pay for the diversity, richness, and refinements of our advanced civilization and our high material standard of living. If our culture were simpler, or if people were more nearly alike, our economy would produce fewer products and could produce each at a lower unit cost. But this is not equivalent to saying that we should simplify our culture or discourage diversity. (It would be a mistake to infer that the majority who play bridge should outlaw canasta so as to reduce the cost of books on how to play bridge. Actually the question whether such restrictions should or should not be imposed takes us outside the field of economics.) Diversity, variety, refinements, richness of opportunity for choice—these are luxury items, like expenditures for curtains and pictures. For an impoverished society it is probably inappropriate to devote a substantial amount of resources to providing these amenities of life while the basic minimum needs of health, comfort, and decency are not yet met. But for a highly industrialized society, capable of meeting its basic needs with only a fraction of its resources, it is not at all inappropriate or inefficient to devote some of the surplus resources to achieving increments of satisfaction through increases in product heterogeneity. Looking at it in this light, that part of the current body of doctrine of welfare economics which rests on the proposition that the minimum point on the average cost curve is the proper criterion of maximization of welfare is not universally applicable. As a generalization it is valid only for the economics of underdeveloped countries.

3. THE QUINTESSENCE OF COMPETITION

In what kind of a market, either real or theoretical, does competition reach its highest degree?

The usual answer is that the ultimate degree of competition is found in the "purely" or "perfectly" competitive market. So far as price competition is concerned this answer is not subject to criticism. Yet such a market is not inconsistent with the complete suppression of quality competition. And where quality competition has been suppressed, competition cannot have reached its ultimate degree since quality differentiation and variation would add to the market's competitiveness by introducing a new dimension of competition.

Of course "pure competition" does not necessarily mean a total lack of quality competition. If an industry consists of a complex of purely competitive markets—that is, if numerous varieties of a product, differing in design or grade, are being sold, each in its own purely competitive market—buyers have a range of qualities to choose from and also enjoy the benefits of pure price competition. But such a situation is not one of unhindered quality competition. In order to keep the products in each submarket homogeneous, individual freedom to offer nonstandard qualities must be curtailed. The ultimate degree of quality competition is reached when every conceivable variety of the product which offers a promise of being best suited to the tastes and needs of some member or segment of the buying population is being offered by some producer, and when innovational changes are being made by the individual producers whenever they are thought to be improvements. If the intensity of quality competition increases with the number of varieties, it cannot have reached its maximum until variation has proceeded to the point where the products of no two firms are alike. But this is "monopolistic" competition; and modern price theory shows that in such a situation the intensity of price competition is diminished by the fact that consumer attachments are not invariably broken by the slightest price differential.

Here, it seems, we are caught in a dilemma. The quintessence of quality competition cannot be achieved unless some departure is permitted from the conditions necessary for the quintessence of price competition—and vice versa.

Yet the conditions of maximum quality competition and maximum price competition are not unconditionally irreconcilable. It is only when we take into consideration the space and population limitations of our world, or, alternatively, the diseconomies of small-scale production, together with the variety of people's basic wants, that we find it impossible to set down conditions under which both price competition and quality competition are maximized. In an imaginary world free from these restrictions, conditions may be found under which perfect price *and* quality competition coexist.

Let us suppose a world in which population within a given area can be increased without limit. This permits us to imagine, say, a hundredfold increase in the size of the present American economy without any corresponding increase in distances, transportation costs, or barriers to communication. The automobile industry could then contain a hundred times as many firms, each employing the same mass-production methods as are being employed today. We might then imagine each of today's cars being manufactured by a hundred competing firms and exchanged in purely competitive markets. Yet this would be most unlikely to happen if competition were free. Even if many competitive producers built cars of the same design, buyers might well have reason to discriminate between them on the basis of differences in quality of materials used, care in workmanship, inspection standards, courtesy, and service. More important, producers who differed in tastes, ways of thinking, and inventive skill would have different ideas concerning desirable modifications that could be made in the design. For a product like the automobile, in which engineering progress and the public's preference for novelty and style changes make periodic model changes advisable, it is difficult to conceive even of a system of oligopolistic markets in which rival firms offered cars identical in quality. And what is true of automobiles is also true, though perhaps less strikingly so, of the great majority of producers' and consumers' goods and services in a modern industrial nation. We must abandon the idea of a system of purely competitive markets developing as numbers increase, and think in terms of a single market containing a growing number of varieties of a heterogeneous product.

An indefinite increase in the number of varieties must sooner or later bring about a situation in which each seller's demand curve becomes perfectly elastic. For if the number of varieties is increased without limit, the differences between them will eventually be forced to become progressively smaller until they reach the point where any two neighboring products are indistinguishable. The market will then consist of a chain of substitutes such that B is indistinguishable from A or C, C is indistinguishable from B or D, etc., yet A and E may be sufficiently distinguishable from each other so that if brands B, C, and D, and F, G, and H, had been nonexistent the producer of E would have had a negatively sloping demand curve. When this state of affairs is reached, each seller's product faces the competition of at least two rival products (it would be two in the one-dimensional case—one on each side of him) which are virtually perfect substitutes, and his control over price has therefore completely evaporated.

Thus we may conceive of an imaginary ideal market, not compartmentalized into a system of purely competitive markets, but rather containing a vast agglomeration of varieties of a heterogeneous product which differ from each other by imperceptible gradations. Viewed as a whole, it is not a market for a homogeneous product, but viewed from the standpoint of any single buyer or seller, each variety faces perfect substitutes. This is the kind of market in which both price and quality competition reach their quintessence. It is truly perfect competition! It is distinguished by the fact that every firm finds itself in a position in which it has no control over the supply or price of any conceivable variety it may wish to produce. Not only does each firm sell its current product under conditions of perfect price competition; there is no possible alteration of the product by which it can escape the rigors of perfect price competition. This kind of competition I shall call *unlimited competition*.

This imaginary world is of course utterly fantastic. Yet the picture it provides of the furthermost limits of competition is a useful one, for it enables us to perceive the characteristics of our real world which bar us from attaining unlimited competition. These characteristics are:

1. The limited size of our world's population, and the barriers of distance. Population cannot be expanded indefinitely within a given geographical area; thus markets are not capable of indefinite expansion, and must necessarily remain limited in size.

2. The diseconomies of small-scale production. If there were no such diseconomies, unlimited competition could be achieved even in a world of limited size. For in the absence of economies of scale long-run costs would be constant; and so long as firms' demand curves retained some downward slope, economic profit would be possible, which would encourage the entry of new firms and varieties, thus forcing existing firms to curtail output and causing their demand curves to become more elastic. This process would continue until so many varieties were produced that each faced one or more virtually perfect substitutes. Presumably this number of varieties would be the same as that which had to be created in our world of infinitely expansible population in order to bring about unlimited competition. The ratio of each firm's output to the industry's output would be the same in both cases, but it would be achieved by wholly different methods: in the one case by providing normal-sized firms with an enormous market; in the other, by reducing firms to miniature size.

3. The diversity and multitude of basic wants. If every buyer who entered a market had precisely the same wants, or belonged to one of a sufficiently small number of groups each composed of persons having identical wants, and if the community's culture were such that the number of wanted activities, and hence constellations of basic wants, were sufficiently small, it would be possible to achieve unlimited competition on this earth even in the face of our limited population, barriers of distance, and diseconomies of small-scale production.

What has just been said may be restated as follows. Competition in the modern world of reality differs from unlimited competition in that significant differences usually exist between closest substitutes. If other things remain the same, these gaps in the chain of substitutes become wider when (*a*) the number of industries in an economy increases, or (*b*) the most economical scale of production in an industry becomes larger, or (*c*) the population, or the per capita demand for an industry's product, becomes smaller. On the other hand, the reverse of any of the changes tends to narrow the gap.

The analysis just given leads to a rather unorthodox conclusion: *Markets for differentiated products differ in character from markets for a single standardized product (if or when they do) only because of the* FEWNESS *of the varieties being offered for sale.* The distinguishing feature of markets to which monopolistic competition analysis is applicable is that buyers differentiate between each variety and its *closest* substitute. It is not the plurality of varieties, or the lack of homogeneity in the market as a whole, but the lack of homogeneity between any preferred variety or small group of varieties and its closest substitute, that defines the situation. If only *more* varieties could be produced, the market would become less imperfect! And if *enough* more could be produced, the imperfection could be eliminated!

This conclusion is precisely the reverse of that of standard price theory, where it is concluded that the imperfection can be remedied only by *reducing* the number of varieties—either to one, or to a small enough number so that each variety can have its own "purely" competitive market.[2]

There are, it seems, two stages of fewness. The first stage is reached when the number of varieties is small enough so that each is significantly different from its closest neighbor. Then we have the phenomenon of product differentiation, and a tipped demand curve. The second stage is reached when the number of firms is so small that each firm must take into consideration the reactions of rival firms. Then we have the familiar phenomena of oligopoly. The traditional analysis attributes only the second set of phenomena to fewness.

[2] See Chamberlin, *Monopolistic Competition*, p. 72 *n*.

4. MAXIMUM COMPETITION IN A LIMITED WORLD

If an actual society cannot hope to attain or even approach the highest conceivable degree of competition, the concept of *unlimited competition* can be of only limited usefulness as a practical standard of evaluation. Is it possible to devise a more practical standard—one which permits the economist to judge a market or economy in the light of the existing conditions with respect to population, wants, and production methods, so that the actual state of competition can be compared with the highest degree of competition *possible under the circumstances?*

This is a difficult order. Since competition is two-dimensional, the intensity of competition must be conceived as a function of two variables: price competition and quality competition. It is easy enough to say that competition as a whole becomes greater if a restructuring of the market intensifies price competition while leaving quality competition unimpaired (or vice versa). The real difficulty arises when we come to consider cases in which price competition can be made greater only at the expense of some loss of quality competition (or vice versa). If it were somehow possible to measure the intensity of price-competitiveness along a scale, and that of quality-competitiveness along another scale, and if the two scales happened to be commensurable or could be made so through some agreed-upon ratio of equivalence, then the highest degree of competition would be that combination of price and quality competition which maximized the area. But can the two kinds of competitiveness be made commensurable?

Another difficulty arises from the fact that competition is dynamic as well as static. Its intensity varies not only with the closeness of competitors but with the fluidity of their positions. With regard to this aspect of competition two points seem worth making:

1. Competition is a process, not simply a state of being. It consists of movement and change, though not necessarily continuous movement. Its essential condition is sufficient fluidity to enable movement to take place whenever an incentive arises. Competition is

thwarted when barriers to movement and change are erected. Maximum competition requires maximum freedom of action.

2. Competition is a social mechanism. Its function is to enable a society to sort out alternatives, select some products and prices, and reject others. In a world of imperfect knowledge its operation involves recurrent experimental changes and the frequent offering of alternatives which afterwards turn out to be inferior. According to this view one measure of competitiveness is the frequency of states of disequilibrium. Another is the quality of the industry's achievements—the satisfactoriness of products and the extent to which costs are reduced and products are improved with the passage of time. The making of improvements depends, in general, on three things: the existence of opportunities for improvements, people's capacities for discovering opportunities, and freedom to act whenever opportunities become known. The first two are unalterable conditions inherent in the environment (unalterable, that is, for the economist, and anyway capable of only slow, gradual change). The third is a variable subject to institutional control. Again, competition is not maximized unless freedom to compete is maximized.

All this suggests the hypothesis that competition is greatest when it is freest. If that is so, maximum competition is that state of affairs which is reached as a result of fully free competition.

The best that can be said with assurance, seemingly, is that maximum competitiveness in any flesh-and-blood economy involves some kind of balance or compromise between intensity of price competition and fullness of quality competition, and that only when freedom of competitive action is maximized can the dynamics of competition operate with full effectiveness. According to this view the competitive ideal is one in which each industry contains, not a limited number of "purely" competitive markets with marked gaps in the chain of substitutes, but rather a congeries of products "spread evenly" over the entire range of desired qualities.

Chapter XIII

INCOMPLETE COMPETITION

IN the preceding chapter we explored the uttermost limits of competitiveness. In this, we shall explore the opposite extreme and shall then attempt to build a scheme of classification that will help us to analyze the competitiveness of markets in which competitive and anticompetitive elements are mixed in various ways and in varying proportions.

1. THE ANTITHESIS OF COMPETITION

According to modern price theory the antithesis of competition is monopoly. And monopoly is defined as the absence of perfect substitutes, or the presence of some degree of uniqueness in a seller's product, or of some control by a firm over its supply. Monopoly is said to be present in a competitive situation whenever a seller's individual product is distinguishable at all from its closest substitutes. Control is thus associated with dissimilarity. Quality differences are the monopolistic element, which is absent only in pure price competition. But monopoly is a variable element, which becomes stronger as the "distance" of a product from its closest substitutes becomes greater, and reaches its ultimate degree only when there are no substitutes whatsoever, and when control over supply becomes absolute.[1]

If we grant the premise of standard theory that pure price competition is the acme of competition, this conception of the antithesis of competition must also be accepted. The antithesis of an infinite number of sellers (of a commodity of given quality) is a single

[1] Chamberlin, *Monopolistic Competition,* pp. 7, 9, 62, 63; Robert Triffin, *Monopolistic Competition and General Equilibrium Theory,* pp. 131–33.

seller (of a commodity of that particular quality); the antithesis of having no control over the supply or price of one's product is having complete control.

But if the acme of competition is conceived to be two-dimensional, embracing quality competition as well as price competition, the standard theorist's view of the antithesis of competition must be rejected. Quality differences and innovational quality changes are essential to the functioning of quality competition. Their absence denotes an incompleteness of competition. Whether or not we choose to call these elements "monopolistic," we most certainly cannot view them as antithetical to competition.

It is true that the existence of quality differences is not quite the same thing as lack of homogeneity of any two firms' products. Strictly speaking, only the latter is monopolistic. In other words, it is not the fact that a product differs from *some* others, but that it differs from *all* others, that constitutes the monopoly element.

Yet even uniqueness is not in itself antithetical to competition. If two or more dissimilar products or qualities or locations are equally advantageous, the "monopoly" of each supplier is a wholly fictitious one and in no way diminishes the intensity of competition. No two athletes in a group ready to start a hundred-yard dash can occupy the same location at the same time—therefore, according to the conceptions of economic theory, each has a monopoly of his location—yet this element of dissimilarity does not affect the competitiveness of the situation. It seems correct to say that when rivals' positions are different yet equivalent, competition is as keen as if the positions were identical. In order that economic monopoly be effective, a producer's position must not only be unique and incapable of being duplicated but must also possess elements of superiority that cannot be matched elsewhere or counterbalanced by other elements of superiority. In the not uncommon market situations in which rival products differ in quality but are roughly equivalent in attractiveness, any "monopoly" elements which exist at any moment are almost wholly irrelevant to the competitive picture.

If the firm's unique position confers substantial advantage upon it, the situation is somewhat different. Yet even this type of situation is not antithetical to competition, unless it has been "frozen" by the imposition of restraints on would-be competitors. Otherwise it provides a challenge and incentive to other firms to overcome their inferior positions by matching or surpassing the firm's performance; it is thus a stimulus rather than a deterrent to competition. An advantage that is easily turned into a disadvantage by competitors' product variation is not well described as a shelter from competition.

In the terminology of modern economic theory fleeting advantages in position in a fluid, dynamic competitive situation are described as monopolistic elements. One cannot dispute the correctness of this terminology so long as it is used consistently. Whether the economist chooses to employ a particular word in a particular sense is more a matter of judgment than of right or wrong. But if "monopoly" is defined as synonymous with "competitive advantage," then it can no longer be claimed that monopoly and competition are antithetical. For in that case monopoly is an essential element in the dynamics of socially useful competition; in fact, every winner of an athletic contest must be termed a monopolist!

What, then, is the antithesis of competition? If competition is viewed as a dynamic process, competitiveness cannot be gauged by the study of firms' *positions* but only by their *conduct through time.* Competition consists of movement and change—of alterations in prices, products, market areas, methods of attempting to reach new customers. In a broader sense it consists not merely of the making of such changes but of the ability and readiness to make them if and when new opportunities or pressures arise. If this is the essence of competition, then its antithesis is the freezing of a situation which would otherwise be fluid and dynamic. It is not uniqueness or superiority *per se,* but the maintenance of an advantageous position through restraints on the freedom of competitors or would-be competitors—in other words, the employment of power to *restrain* or *block* or *suppress* or *obstruct* or *exclude* competition that would otherwise occur. This power is not a prerequisite of product differentiation or of vigorous quality competition. It is a force which, when exercised, works against both price competition and quality competition.

Competition can also be regarded as a social process by which alternatives are made available and those rated most highly are selected. If competition is viewed in this light, its antithesis is, again, the restraint of competitors' actions. For such restraint represents the suppression of alternatives, or the denial of opportunities for free choice. Buyers' freedom of choice is limited or destroyed when sellers agree to combine, as in a cartel, or when one seller manages to destroy all his competitors, as in a true monopoly, or when the government intervenes in order to restrict variations in the product, or limit quantities, or fix prices. Likewise sellers' freedom of choice can be limited or destroyed by the collusion or coalition of buyers or by government intervention. Such actions restrict individual freedoms unequally, since they provide freedom to one contestant to do what he wants while denying it to another. It is this aspect of interference with competitive freedom that distorts or paralyzes the functioning of the competitive mechanism. According to this view, the nadir of competition is reached when freedom of choice becomes zero.

The anti-competitive force may be appropriately described by such terms as *interference, suppression,* and *restraint* of trade. It can also be described by the word *monopoly* if that word is used in its popular sense. The type of monopoly which inspired the American antitrust laws, which was represented in the old Standard Oil Company and the chartered companies of Elizabethan England, and which has restrained competitors from running a second bus line down Fifth Avenue, is more than mere possession of a product or process or location unlike anyone else's; it is a control over supply attained by forcible blocking of man's freedom to enter a market.

One idea which seems pervasive in modern price theory is that solitary occupancy of some "location" or singleness in the offering of some variety of product is equivalent to control over its supply. Actually, to be the sole source of supply of anything and to have control over that supply are two different things. This becomes evident if we broaden our vision to include the time dimension. The word "control" implies some degree of mastery over the future course of events, not simply a position in one instant of eternity. Firemen have a fire "under control" only when they can prevent it from spreading. The test of whether a driver has his car "under control" is not its momentary position on the road but his ability to keep it on the road; if the next dip or curve in the road might send it into the ditch, and if he is powerless to prevent that occurrence, it would hardly be appropriate to say that he had his car under control. Similarly, "control over the supply of a commodity" carries some implication of power to govern the future supplies of that commodity. If monopoly is defined as sole possession of some source of supply, then two kinds of monopoly need to be distinguished: (1) unprotected monopoly, or monopoly by temporary possession—which describes the position of a firm which happens to be at the moment alone in its field, but lacks the power to prevent others from moving in if they so choose; and (2) protected monopoly, or monopoly by restraint, in which a continuation of the present position of isolation is assured by the existence of effective institutional barriers to entry. If monopoly is defined as "control over the supply," only the second type of situation can be called monopoly.

Sometimes the phrases "sole possession" and "exclusive possession" are used as if they were synonymous. If the word "exclusive" is interpreted literally, however, it denotes exercise of the power to exclude others who want to enter, and therefore "exclusive possession" is appropriate only to cases of protected monopoly whereas "sole possession" is a more general term, applicable both to the unprotected innovator and to the protected patent-holder.

The distinction is an important one for that branch of economics which seeks to offer guidance in questions of public policy. It demarcates the transient position of advantage from the bulwarked position of the institutional monopolist. It is true that the unprotected position includes the "natural monopoly" situation, in which a firm retains a more-than-ordinarily profitable position without erecting artificial barriers simply because other firms do not find it worth while to enter. Price theory has performed a service in showing that positions of monopoly and exploitation which diminish welfare need not rest on legal foundations. But in employing a single blanket term to describe both temporary differentials in position in a dynamic competitive situation and monopoly positions which can be retained only by coercive methods or grants of privilege it may have performed an even greater disservice, for it has succeeded in obscuring the vast gulf of difference between these two kinds of position. When seeking ways to improve the organization of society, it seems desirable to distinguish, say, between the nature-made havoc wrought by floods and hurricanes and the man-made havoc wrought by wars, riots, and sabotage. Likewise it is useful to distinguish between deviations from a conceptual ideal which are ineradicable from a competitive private enterprise society and those which have their roots in the particular system of laws and government policy which exists in a given nation at a given time.[2] Even more, it is important for the welfare economist to be able to distinguish between striving to excel others and striving to prevent others from excelling. The business firm which maintains its established position by dint of special privileges or threats that deny others the opportunity to serve the community's interests better is no more socially constructive than the shopkeeper who wrecks his competitor's establishment. And it is the polar opposite of the business firm which owes its success to its ability to produce something better than the best products hitherto available, or to produce the same thing cheaper. It is no less important in economic affairs than in amateur athletics to differentiate between the contestant who eliminates others before the contest begins and the contestant who eliminates others as a result of superiority in the contest.

Talcott Parsons makes this same point when he distinguishes "two great classes or modes of acquisition": one in which A acquires what he wants from B by widening B's otherwise existing range of acceptable choice (B being the customer or judge), and the other in which A gets what he wants from B by narrowing B's alternative courses of action or making them more difficult or unpleasant. This

[2] See Friedrich A. Hayek, *Individualism and Economic Order*, p. 105. His distinction is between "the prevention of competition" and "the shortcomings of actual competition compared with an ideal which is irrelevant for the given facts."

latter mode of acquisition, it should be noted, has unpleasant consequences not only for B but for those others who are also seeking to acquire what they want from B.[3]

The distinction just made between protected and unprotected monopoly may be combined with the distinction made in Chapter VII between single-variety or brand monopoly and industry-wide monopoly,[4] to provide an interesting cross-classification of market situations. This is shown in Table 1.

Table 1. Four Types of Solitary Position

	I (Aloneness with respect to the whole industry)	B (Aloneness with respect to a single brand)
P (Protected by power to exclude)	PI Effective barriers against entry of newcomers into the industry. *Complete monopoly.*	PB Effective barriers against imitation of a product's particular qualities. *Monopolistic competition.*
U (Unprotected)	UI One firm the only occupant of an industry. *Temporary monopoly, with potential competition.*	UB Products dissimilar, but open to imitation. *Complete competition.*

One thing to notice is that Table 1, although confined to situations containing "monopoly elements," includes the fullest kind of competition as well as its polar opposite. The presence of brand monopoly, *if unprotected,* or, alternatively, of unprotected monopoly, *if limited to a single brand,* does not of itself diminish an industry's competitiveness. True, brand monopoly creates attachments that are not easily broken. In industries composed of branded products, and especially those characterized by typical modern capitalist institutions and consumer attitudes, new brands cannot win immediate acceptance. But this is merely an instance of the

[3] See Talcott Parsons, "Some Reflections on 'The Nature and Significance of Economics,'" *Quarterly Journal of Economics,* XLVIII (May, 1934), 527; and also Nicholas Kaldor, "Professor Chamberlin on Monopolistic and Imperfect Competition," *Quarterly Journal of Economics,* LII (May, 1938), pp. 524–25.

[4] Above, pp. 85–88.

universal law that, because of the costliness of attaining knowledge, untried things and ideas are at a disadvantage in competing with established things and ideas. When competing products are dissimilar, identifying brand labels are a prerequisite of efficient quality competition. If brand names were abolished in an attempt to eliminate the monopoloid elements of good will and brand preferences, quality competition would be substantially lessened.

The two elements which *do* diminish competitiveness are (1) the "I" element: the absence or elimination of competing alternatives, or, we might say, the expansion of the area of aloneness until it includes the whole industry; and (2) the "P" element: the prevention of future change in the number or character of the alternatives—the freezing of the existing situation. When these two elements are combined, the suppression of competition is complete.

Another thing to notice is the difference between the situations UB and PB. Situation UB is characteristic of thoroughly unhampered competition. Situation PB differs in that it supposes protection against encroachments; the firm has its own impregnable position. If a whole industry were composed of firms enjoying such protected positions, each firm would have a "local" monopoly yet face the competition of less satisfactory substitutes. This is the situation supposed in the theory of monopolistic competition.[5] It is obvious that behavior in the two types of situation may differ markedly, and that a theory appropriate to the PB type of market may be thoroughly inappropriate as a tool for interpreting behavior or the state of welfare in markets of the UB type.

Still another thing to note is that mere absence of competition is not the polar opposite of complete competition. This may seem paradoxical until it is realized that the time element is important. Competition is a process which serves a specific social purpose: that of altering the existing alternatives whenever some contestants believe that new and better products or terms can be offered. In UB situations competition may be inactive some of the time. But the fact that competitive forces are ready at such times to come into play the moment the occasion warrants their reactivation makes the

[5] See Chamberlin, *Monopolistic Competition*, p. 200: "With respect to the *particular product* produced by an individual firm under monopolistic competition . . . no one else can produce a product identical with it."

situation significantly different from that of a protected cartel. Similarly, a UI situation is significantly different from a PI situation in that potential competition stands perpetually ready to turn into actual competition. It is also different in that it is probably an exceedingly rare phenomenon. It is to be found in a few public utility situations, in even fewer new industries (for example, the first commercial broadcasting station during the brief interval before the second appeared), and in occasional situations of geographical isolation—though geographical barriers seldom create positions of total monopoly. Modern economic history seems to show that sustained absence of competition is highly correlated with the suppression of competition.

Thus, according to our analysis, the real antithesis of competition is not monopoly in the sense of sole possession but coercive limitation or elimination of alternatives—that is to say, monopoly in the sense of present and future control over the supply.

Two things are worth observing about the analysis just completed. First, the notions of complete and prolonged absence of competition within an industry as the polar opposite of competition, and of interference with competitors' freedom of movement, rather than isolation of position, as the anti-competitive force, rest on the analytical structure developed in earlier chapters: the concepts of constellations of wants, of the industry, of the two-dimensional character of competition, and of competition as being a dynamic process. And second, these notions, while in conflict with those of present-day price theory, accord with the traditional economic, political, and legal views that prevailed until the advent of monopolistic competition theory. As Wallace has expressed it, "The anti-trust laws did not . . . make monopoly illegal. Rather, the thing condemned was monopolizing or restraint of trade . . . by . . . combination . . . or by destruction or weakening of existing competitors or blocking of potential competition by the use of bludgeoning, harassing, or obstructive tactics rather than competitive methods which measured relative efficiency in production and marketing." [6] Legislative circles continue to hold this view. According to the House Judiciary Committee of the 81st Congress, for

[6] Donald H. Wallace, "Monopolistic Competition and Public Policy," *American Economic Review: Papers and Proceedings*, XXVI (March, 1936), 79–80.

example, antitrust laws are directed against "combination, conspiracy, or collusive agreement, or monopolistic, oppressive, deceptive or fraudulent practices" [7]—in other words, positive anti-competitive acts.

From the standpoint of policy to maintain competition in a free individualist economy the primary problem today is not monopoly but interference—which, if carried far enough, may lead to monopoly, yet is deleterious to welfare even when it is not carried that far, but takes the milder form of collusive agreements to fix prices or quotas, or abstain from quality improvements, or keep newcomers out of an industry, or of tacit understandings to refrain from competitive behavior, or of pressure-group-sponsored government measures to prevent or restrict competition. Probably the biggest part of this problem is oligopoly. Limitations of space (and on the reader's patience!) prevent more than mere mention of this problem area: the conditions under which fewness or bigness leads to results not significantly different from those of oneness. But here, too, the "monopoly" or "dissimilarity" element is almost wholly irrelevant to the problem; the objectionable features of the situation are largely those caused by interference with independent competitive action.

It is a relatively minor matter, of course, what terminology is used—whether the anti-competitive force is labeled *interference* or *monopoly by restraint* or simply *monopoly*—provided that the needed concepts and distinctions are made clear. We might define monopoly as "selling alone." In that case temporary positions of competitive advantage or isolation could be described by such terms as *unprotected monopoly* or *temporary monopoly* while positions maintained by obstructive methods could be described as *protected monopoly* or *monopoly by restraint* or *monopoly by exclusion*. Or we might define monopoly as "exclusive control." This would limit the term to situations in which the seller can maintain his aloneness. Temporary positions would then be described by such terms as *competitive advantage* and *isolated selling with potential competition*, while protected positions would be fully described by the single word *monopoly*.

[7] *New York Times*, March 15, 1950.

It is not within the province of any one writer to decide which concept of monopoly economic theorists shall adopt. But whichever is adopted, it should be employed consistently. To identify monopoly sometimes with singleness ("selling alone") and sometimes with exclusiveness ("control"), as if the two were identical, is the error that needs to be avoided.

2. A CLASSIFICATION OF MARKETS

From time to time economic theorists have proposed classifications of types of markets according to their degree of competitiveness, employing the current theory of monopolistic competition as a guide.[8] These classifications have differed in detail, but all have been in agreement on one point: pure competition belongs at one end of the scale, pure monopoly at the other, and monopolistic competition somewhere in between.

But if the premises of monopolistic competition theory are replaced by the propositions developed in this book—that competition is two-dimensional and dynamic and that the anti-competitive force is interference with freedom of action and range of choice—a quite different kind of classification must be devised. For it then becomes logical to classify markets according to whether competition is permitted to operate freely or is restrained by some form of interference. At one end of the spectrum can be placed the wholly free markets (complete competition), and at the other end the wholly controlled markets (complete suppression of competition), while in between can be arranged the various market situations in which interference occurs only in certain respects, or in certain portions of the market. Probably the most important type of partial competition is that in which one dimension of competition is unhindered while the other is restrained. That is, we may find either price restraints or quality restraints or both.

The simplest scheme is a three-level classification containing four types of markets:

[8] See esp. Robert Triffin, *Monopolistic Competition and General Equilibrium Theory*, pp. 104, 143, and Fritz Machlup, "Monopoly and Competition; a Classification of Market Positions," *American Economic Review*, XXVII (Sept., 1937), 445–51.

TOP LEVEL: Complete competition
INTERMEDIATE LEVEL: Pure price competition
Pure quality competition
BOTTOM LEVEL: Absence of competition

A more elaborate scheme can be devised. taking into account not only the kinds of competitive freedom permitted but also the number of firms participating, which sets an upper limit to the possible number of independently offered dissimilar alternatives. Such a scheme is shown in Table 2. It is somewhat similar to Table 1 in

Table 2. A Classification of Markets

	Monopoly (isolated firm)	Oligopoly (second-stage fewness)	Polypoly (first-stage fewness)	Infinopoly (unlimited numbers)
Complete restrictions	Complete monopoly	Cartel	Cartel	Cartel
Restrictions on pricing	Potential quality competition	Oligopolistic pure quality competition	Polypolistic pure quality competition	Unlimited quality competition
Restrictions on quality	Potential price competition	Oligopolistic pure price competition monopolistic competition	Polypolistic pure price competition monopolistic competition	Unlimited price competition
No restrictions	Potential competition	Complete oligopolistic competition	Complete polypolistic competition	Unlimited competition

the previous section, differing from it in that it takes into account (*a*) restrictions on price as well as on product, and (*b*) fewness as well as aloneness. The phrase "first-stage fewness" refers to numbers which, though large, are not large enough to eliminate significant gaps between closest substitutes. "Second-stage fewness" refers to numbers sufficiently small to cause competitors to take into consideration their rivals' reactions.

One type of market included in this classification is *monopolistic competition*. The term is used here, not in its ordinary sense, but to define an industry composed of protected brand monopolies in competition with each other. Such a situation differs from *complete competition,* which stipulates full freedom to imitate rival qualities, and also from *pure price competition,* which bans not only freedom to vary quality at will but also all heterogeneity. The place assigned in Table 2 for this type of market is not wholly satisfactory. In one respect such a market occupies an intermediate position between the fully competitive market and the one which lacks quality competition altogether. But in another and possibly more important respect it is less competitive than either, for by permitting the monopolization of existing variants it restricts new firms to inferior positions.

Our classification has other drawbacks too. It does not distinguish gradations of interference. It does not distinguish partial from total interference in any given area of competition. It provides no place for industries which are completely competitive for those already in them, yet prohibit the entry of newcomers. It does not distinguish temporary acts of interference aimed at cushioning the impact of a change, and modifying or retarding an adjustment, from lasting restraints which prevent adjustment. It does not enter into the question of motive or purpose, and so does not distinguish intervention for the sake of stability, or security, from interference for the sake of unwarranted private gain.

Perhaps these defects could be remedied in a still more elaborate scheme of classification. Yet the classification as it stands is useful. It provides a practical yardstick by which the competitiveness of markets can be appraised. The kind of competition which is held up as a standard is not impossible of attainment. The modern business world contains numerous industries in which the actual market situation has at times come fairly close to the ideal—though, of course, numbers of competing varieties being limited, the result has not resembled unlimited competition. Our scheme thus offers, with respect to those markets in which unhindered competition is held to be desirable, a usable guide to public policy.

It also has an advantage over *ad hoc* yardsticks in that it is based not only on a "realistic" concept of competition but also on a theory which is consistent with that concept and in fact supports it. It thus merits greater confidence than a conceptual yardstick which is either theoretically derived but unworkable, or workable but at odds with theory.

Chapter XIV

SOME CONCLUSIONS ARE DRAWN

HOW useful and how important is the theory that has been developed in this book? Does it throw any new or different light on economic situations and behavior typical of modern industrial societies? Does it furnish a more adequate explanation of observable economic conduct than does the currently accepted Theory of Imperfect Competition? Does it materially alter our criteria for judging the "social efficiency" of various types of business behavior and market structures? Does it offer a more reliable guide to public policy?

No author is in a position to appraise his own work coolly and objectively. Yet these questions cannot be ignored. If we are going to face them, it may be helpful to look back over the road we have traveled and attempt at least a provisional evaluation of what this essay in theory has accomplished.

1. AS TO THE THEORY OF WANTS AND CHOICE OF QUALITY

The analysis presented in Part Two, relating consumers' and producers' choice of quality to basic wants, has enabled us to construct a more general theory of economic behavior, and one in which sharper distinctions can be drawn, than conventional analysis makes possible.

The theory rests on a number of explicit assumptions concerning the nature of our world and of human conduct, including two which other social sciences consider vitally important, yet which economic theory customarily abstracts from: (1) the diversity and changeability of people's circumstances and tastes, and (2) the imperfection of knowledge, and hence the importance of experimentation and the learning process. An equally important foundation stone is the premise that wants are desires for experiences, which products ordinarily satisfy only imperfectly.

These assumptions enable us to penetrate the area that lies behind the indifference maps of present-day demand theory. Instead of merely accepting these maps as data, the theory conceives them as being derived from more fundamental data. It thus embraces what J. M. Clark has called "the dynamic process of preference formation," [1] usually omitted from the theory of competitive markets. It also embraces the dynamic process of quality variation, which is viewed not as something exogenous, nor as an element of imperfection, but as one aspect of the movement toward equilibrium generated under conditions of wholly unfettered competition. It provides room for analysis both of adjustments made on the basis of the existing set of commodities (as does the standard modern theory) and of adjustments in the make-up of the commodities themselves on the basis of given consumers' aims.

The theory leads to several conclusions that are at variance with those of standard modern theory:

1. Socially useful entrepreneurial activity consists not only of activity dealing with production methods, costs, outputs, and prices, but also of (*a*) activity that seeks to achieve greater precision in satisfying wants in connection with the sale of existing products, by altering the knowledge and attitudes of buyers, and (*b*) activity that seeks to achieve greater precision by altering the qualities of products.

2. Product heterogeneity, product variation, and advertising are means of attaining a competitive equilibrium otherwise impossible. It is erroneous and misleading to treat these phenomena solely as elements of imperfection in the competitive process. They may be so under some circumstances, but an analysis which overlooks their positive contributions to competitive equilibrium and social wel-

[1] John Maurice Clark, "Realism and Relevance in the Theory of Demand," *Journal of Political Economy*, LIV (Aug., 1946), 349.

fare is seriously incomplete. They are needed in any rational exchange economy.

3. Advertising is not an identical alternative to price reduction; it performs a different function. A price reduction is aimed at those who are already familiar with the product but dissatisfied with the price. Advertising is aimed at a different group of people: those who are unfamiliar, or not sufficiently familiar, with the product.

4. The outcome of full freedom to compete is not as a rule perfect or pure competition. Except in rare instances, therefore, the conditions of pure competition cannot be realized without institutional restraints on entrepreneurial freedom to vary the product, even in the absence of behavior aimed at monopolizing the market or hoodwinking customers. The assumption of a homogeneous product in competitive price theory thus seems to be inconsistent with the assumption of freedom of competition.

5. The conditions of so-called "pure competition" cannot, in general, maximize welfare.

6. Competitive markets generate processes which, in addition to determining the qualities and prices of products and the methods of producing them, necessarily result in economic development.

This last statement points to an important respect in which the theory developed in this book is more general than the theory now current. By showing how the very process of exchange gives rise to evolutionary change it unites the theory of competitive markets with the theory of economic development, forming a single system. It thus explains economic development not as a spontaneous disturbance that upsets the system's equilibrium but as part of the general striving toward equilibrium. This contemporary equilibrium theory has failed to do, chiefly because it has been exclusively a theory of price and so has limited itself to the case of an economy producing n commodities of constant quality.

The theory is more general in still another way. It embraces both the socially constructive and the socially damaging elements in non-price competition, whereas formal theory today almost completely obscures the constructive element. It thus becomes possible for the theoretician to examine the conditions under which each element is most likely to predominate.

The analysis also has the advantage of disclosing behavior patterns unrevealed by current theory which are commonplace and in fact pervasive in the actual world. A substantial amount of economic activity in most markets is motivated by buyers' concern with products' qualities, uses, variety, novelty, improvement, deterioration, etc., and producers' concern with market research, quality control, sales policies, and other marketing problems involved in reaching new customers and serving old ones better. Such activity cannot be predicted or explained by the apples-and-nuts analysis of modern demand theory or the cost-price-output calculus of the theory of the firm—in fact standard modern theory inevitably makes such activity appear superfluous or even mischievous—whereas it is actually (in a world of diverse tastes and imperfect knowledge) an essential part of maximizing behavior.

E. E. Schattschneider has drawn attention to an analogous shortcoming in the political theory prevalent in the predemocratic era. The theorists of that time took it for granted that popular sovereignty would be effective automatically in a democracy; they therefore failed to see the necessity for political parties or to predict their establishment, having excluded them by hypothesis.[2] Similarly, formal economic theory has been built on the assumption that wants become translated into demands automatically, effortlessly, and perfectly, and so has failed to provide a place in its "purely competitive" system for the elaborate institutions which are actually required to effect the translations.

In this regard modern theory seems deficient in two respects. First, in not having advanced beyond the oversimple first approximation in which individuals are omniscient and homogeneous. And second, in being limited to the quantitative aspects of economic decision-making, and consequently leaving the qualitative aspects impounded in *ceteris paribus* assumptions which reduce society to a stationary state in which products' qualities and people's attitudes toward them remain constant. These assumptions are of course useful and in fact necessary for analysis of equilibrium price determination, but for analysis of the market process as a whole they assume away the very things the analysis ought to reveal.

[2] E. E. Schattschneider, *Party Government*, pp. 13–15.

2. AS TO THE ANALYSIS OF QUALITY COMPETITION

In Part Three a theoretical framework was established in which two dimensions of competition are recognized: quality competition and price competition. It was shown that there are two "pure" forms of competition: (1) "pure price competition," in which quality is uniform and unchanging, and competitive adjustments are restricted to alterations in price, and (2) "pure quality competition," in which the price is uniform and unchanging, while quality is freely variable, so that competitive adjustments are made solely through the medium of qualitative changes in the competing products.

Analysis of pure quality competition revealed that the results of a rigid price situation may be quite different from those which conventional price analysis would lead us to predict. In fact, quality competition performs, in the absence of price competition, much the same functions that price competition performs.

It was shown too that quality competition performs other functions which cannot be performed by price competition. It enables buyers to satisfy their wants with greater precision. It provides a range of choice broad enough to satisfy buyers who differ widely in their circumstances and tastes. It promotes novelty, variety, and progress.

Our analysis also revealed that the type of market in which competitive firms offer products whose qualities are dissimilar, freely variable, and imitable, is actually *more* competitive than the so-called purely competitive market. It is more competitive because it contains a second dimension of competition—quality competition—which is necessarily either absent or restricted in the case of the "purely competitive" market.

This conclusion is, of course, exactly the reverse of that reached by Chamberlinian monopolistic competition theory. The reason why the latter theory views such a market as *less* competitive is that it is a branch of price theory. Its focus is on price determination, and its concern with quality differences and changes is almost wholly a concern with the effect of these factors upon prices. It is easy to see why, under the circumstances, economic theorists have, with only a few exceptions, failed to regard quality differences and changes as competitive rather than anti-competitive factors.

One conclusion seems inescapable: the state of economic welfare cannot be described correctly if attention is focused solely on prices. Welfare economics and socialist economics draw erroneous, or at least incompletely justifiable, inferences from economic theory to the extent that their conclusions rest on the assumption implicit in a purely quantitative analysis that the existing kinds and qualities of goods are to be taken for granted as determined (in some automatic way) by existing tastes and technology. The truth is that they cannot be taken for granted. They are dependent to at least some degree on producers' freedom to vary quality and innovate, and on incentives for so doing.

How important is quality competition? It is tempting, but unscientific, to hazard a guess that quality competition has been every bit as important as price competition, perhaps even more so. It is certainly the great whetstone of material progress, and has played a leading role in the development of our amazing modern industrial civilization. This, however, is not the most important conclusion to be drawn. What is crucially important is this: Only when quality competition is operating can a competitive individualist economy realize its full potentialities.

Three other points seem worth stressing:

1. Effective quality competition requires fewer numbers than effective price competition. Price theory tells us that a duopoly in the sale of a homogeneous product is hardly better than a monopoly, and that oligopoly lies far closer to monopoly than to pure price competition between many sellers. But this is not true of quality competition. Two dissimilar varieties are vastly better than one, provided quality is variable, and buyers are alert to differences and changes. And a relatively small number of varieties is almost as good as many. It is worth noting that in democratic politics, which employs pure quality competition in the selection of leaders (except in those occasional elections in which money changes hands), Western democratic societies consider the two-party system (duopoly) infinitely superior to the one-party system (monopoly).

2. A uniform, unchanging price is not necessarily evidence of lack of competition. Absence of price variation indicates merely that competition—if there is any—is restricted to the quality dimension. Since the equivalent of price flexibility can exist in markets characterized by complete uniformity and rigidity of prices, it is only when quality is also uniform and rigid that competition is lacking. Unless economic investigators are aware of the inverted "price" competition that can occur when vertical quality is variable, they may be led into serious error in their appraisal of the competitiveness of certain industries.

In industries characterized by quality flexibility, the term *price rigidity* is inappropriate, since it has no economic significance. The term is surely misused when applied to situations in which the price of B on Tuesday is the same as the price of A on Monday; it can be meaningful only when both prices refer to the same product, and to a specified unit of specified quality. When the unit is fluctuating in size or quality, we are moored to a shifting anchor; price would have to fluctuate similarly in order to produce a situation comparable to that of a rigid-price, rigid-quality market.[3]

3. Quality differences and quality variation are aids to price competition in oligopolistic markets. When competitors are limited in numbers and their products are identical, the pitiless glare of pure price competition can be unbearable. The slightest price difference becomes intolerable. Under these circumstances price competition is unworkable. Competitors tend to seek relief in collusion or tacit agreement or a live-and-let-live attitude. Quality differences can ease such a situation by creating a mist of uncertainty behind which moderate price differences may arise and be varied. This tends to bring about a situation of greater price fluidity.[4]

[3] Cf. Joseph A. Schumpeter, *Capitalism, Socialism, and Democracy*, p. 92.

[4] Cf. John Maurice Clark, "Toward a Concept of Workable Competition," *American Economic Review*, XXX (June, 1940), 249; Robert Triffin, *Monopolistic Competition and General Equilibrium Theory*, pp. 75, 86; M. A. Adelman, "Effective Competition and the Antitrust Laws," *Harvard Law Review*, LXI (Sept., 1948), 1299–1300. Incidentally, it is possible that the rather puzzling entrepreneurial behavior in the cement industry, in which differences of quality between competing brands of "standard Portland" cement have been played down by producers when competitive advantage might have been gained by playing them up, can be explained as due to a desire on the part of members of that industry to avoid the rigors of competition coupled with a realization that avoidance of competition becomes much more difficult when the product loses its homogeneity.

If our analysis of the operation and effects of quality competition has been correct, a reconsideration of the oligopoly problem and a reappraisal of the competitiveness of the American economy seem to be in order.

The view is widely held today that (1) most markets in the United States are oligopolistic, (2) oligopolistic markets are usually characterized by price inflexibility or stickiness, (3) the degree of inflexibility or stickiness indicates the degree to which competition has become moribund, and therefore (4) competition is largely inoperative in the United States today. The conclusion, it must be noted, rests on three premises. The first two premises may well be correct; but if the third is incorrect, the conclusion does not follow. Analysis of the operation and effects of quality competition reveals a dimension of competition independent of price inflexibility. If this is taken into consideration in an empirical study of the competitiveness of the U. S. economy, the currently accepted conclusion may need to be modified considerably.

Let us take a single example of an oligopolistic market: the fiction magazine industry. If this industry is studied from the economic viewpoint generally held today, it is described as one consisting of a few large-scale firms whose products are branded, differentiated, and sold at administered prices which remain rigid and identical for long periods of time. What "competition" occurs takes the form of nonprice competition: incurring of selling costs, solicitation of subscriptions, newsstand displays, and product differentiation and variation. In short, here are all the phenomena usually associated with an oligopoly stalemate.

But if we adopt the approach suggested by the foregoing analysis of consumer behavior and quality competition, a different description emerges. Firms which publish fiction magazines exist in order to satisfy basic wants. People desire to be entertained, distracted, informed, and transported into imaginary worlds. One of the most important of the entrepreneurial functions of a magazine publishing firm is that undertaken by the editor. His job entails a day-to-day search for ideas, manuscripts, departments, special features. His is the difficult task of matching the desires of a public whose tastes must be guessed at with a product that needs to be designed anew every week. The aim must always be to satisfy the public's wants a little more accurately than any previous issue did, and than any competing magazine is doing. Here price rigidity is only nominal. With each issue a different bundle of utilities is offered. If competition presses hard, efforts must be made to heighten the attractiveness of each issue; better stories or more famous authors must be secured, necessitating higher costs of production. Other alternatives are open also, of course. The newsstand price or subscription price can be cut, or advertising rates reduced. But these are likely to be less powerful levers to competitive advantage. According to this view, the oligopoly situation is anything but a stalemate.

All this suggests the need for a broader approach to the oligopoly question. Unless quality is uniform and invariant (and actual prices coincide with published prices) the price theorist's analysis of oligopoly is based on incomplete data, and is therefore inadequate. This does not mean that the problem of price inflexibility is a myth, or is easily solved, but merely that it may be less of a problem than conventional theory has led us to believe.

Our analysis also has implications for public policy with regard to those oligopolistic markets in which both price competition and quality competition are sluggish. It suggests the hypothesis that the best way to activate competition is not to try to stimulate price competition directly—this may be an almost hopeless task, especially if there is collusion or quasi-agreement—but rather to encourage both quality differentiation and quality variability. For when quality differences arise, the deterrents to price changes are weakened; and when quality keeps changing, whatever price inflexibility remains loses its economic significance.

3. AS TO THE DISTINCTION BETWEEN MONOPOLIZING AND COMPETITIVE ADVANTAGE

According to our analysis an important element in producers' behavior is their competitive search for products that best fit the public's constellations of basic wants. The product variation which ensues, largely on an experimental basis, is an essential, socially desirable activity. If an economy is to give its best performance, such activity must be encouraged.

Quite different is that form of producers' behavior whose aim is the exclusion or suppression of rival products in an industry or the placing of competitors at a cost or quality disadvantage through legal or other restrictions on the freedom of other firms to imitate successful products or production methods to whatever degree it is technically possible for them to do so. Here the action is a socially undesirable one, causing a diminution of consumers' welfare.

It is important, if analysis is to assist policy making, that categories be selected which enable us to distinguish between activities that contribute to social net product—and corresponding returns that are a measure of efficiency and merit—and those which do not. The analysis in this book permits such a distinction to be made, whereas it is blurred in the monopoly analysis provided by modern price theory. There no distinction is made between the "monopolistic" gains of the constructive business pioneer whose rivals have not yet caught up with him and the gains very different in character extracted by the "racketeering" device of erecting roadblocks in the paths of competitors. The first is a relatively insecure position of competitive advantage; the other is a lasting position, sheltered from the rigors of competition. The first reflects a betterment which not only produces an immediate gain in welfare but encourages others to make further gains; the other is a "worsement" (relative to the welfare position under unhindered competition) and an obstruction to future betterments. Modern price theory, almost completely preoccupied with static equilibrium positions, has veiled this important distinction by assuming away the state of flux and uncertainty normal to the real world, so that all unique quality differences are regarded as necessarily permanent and inimitable. This narrows the concepts of disequilibrium and adaptation to a purely quantitative basis and forces the theorist to classify all differential returns other than windfall and adaptation profits as calculable rents, imputable to factors other than "enterprise."

Our theory, on the other hand, by assuming explicitly uncertainty about utility and broadening the concept of adaptation of means to given ends to include changes in the quality of products and in production methods, makes possible the useful distinction between the calculable, persistent elements in business income, such as those due to a monopoly franchise, and the incalculable, fleeting elements due to temporary superiority gained by real or illusory quality improvement or cost reduction.

This distinction, together with the distinction between brand monopoly and industry-wide monopoly developed in Chapter VII, provides useful tools of analysis for the economist concerned with antitrust policy which hitherto have been lacking in theoretical economics.

4. FINAL CONCLUSIONS

If the analysis in this book has been correct—if there is such a thing as quality competition, distinguishable from price competition—two vitally significant conclusions emerge. First, competition between homogeneous products is not the highest form of competition. And second, quality differences and changes are not, in general, a monopolistic imperfection but a competitive element. Instead of choosing homogeneity and heterogeneity as the two polar extremities of competition, we must adopt complete competition (which permits heterogeneity of both product and price) as one extremity and complete interference (which may result in homogeneity of both) as the other.

No theory of quality determination and quality competition can pretend to be a substitute for price theory. It can only be a supplement to it. As Chapter XII has made clear, it is merely the "other half" of a complete theory of competition. Not until both halves are pieced together can a full analysis of competitive forces in a private-enterprise economy be achieved.

To this element of incompleteness must be added another. Our theory deals only with economic conduct that occurs when consumers are to some substantial degree successful in forming and maintaining their own tastes and values, in informing themselves about products, in making use of their reasoning powers, and in avoiding being victimized. If producers can in fact create basic wants and manipulate value systems virtually at will, or if rational calculation of the best means to given ends is an insignificant element in consumers' choice, or if consumers' "information" bears but little resemblance to the objective facts, a different sort of theory would be needed to explain predictable behavior. However, so long as deviations from rational-ethical conduct do not entirely destroy the reality of ends or the ability to act rationally in selecting means or the possibility of acquiring knowledge, our theory is useful, both as a means to prediction (though an imperfect one) and as a guide to improvement in welfare.

It must be acknowledged that the framework of analysis presented in this book fits some markets and some buying situations within a market better than others. It is most appropriate to markets in which the element of novelty looms large, and to the household's first few purchases of a product. One can think of a number of commodities (standard hardware items, for example) to which it hardly applies at all. Actually, the degree to which purchases are experimental, investigation is needed, a range of alternatives is beneficial, or quality improvement is desirable, varies from market to market and even from purchase to purchase, depending upon particular circumstances.

In order to obtain a general picture of how markets operate in an exchange economy the reader may find it helpful to think of the various kinds of products as capable of arrangement in an array. At one extreme are the goods that are completely standardized, technically incapable of being varied in quality, divisible into small units, constantly bought, and perfectly familiar to buyers. At the other end are those goods that wholly elude standardization, are variable in quality, indivisible, bought no more than once in a lifetime, and totally unfamiliar to buyers. Actually these extremities remain unpopulated; they are merely theoretical limits. The goods in our world of reality are scattered at various places between the two extremes.

This analytical device may help to make clear the very great difference between present-day price theory and the theory of two-dimensional competition presented in these pages. Competitive price theory views only the first extremity, and is applicable only to this limiting and unreal case. The theory developed in this book, on the other hand, is concerned with the entire spectrum. For convenience in exposition, the reader's attention has been focused on those products that lie near the second extremity, yet actually the theory is applicable to all points on the spectrum with the single exception of the aforementioned unattainable extremity at which knowledge is perfect and variability impossible. In this respect the theory presented here is, as the other is not, a *general* theory of competitive markets.

Bibliography

The following is not intended to be a comprehensive list of works relevant to the subject matter in hand, but simply a list of the books and articles to which footnote references have been made in this book. Its chief purpose is to serve as a guide to the reader in identifying sources where full identification is not given in the footnotes.

Adelman, M. A. "Effective Competition and the Antitrust Laws," *Harvard Law Review,* LXI, Sept., 1948.

Ayres, C. E. *The Theory of Economic Progress.* Chapel Hill, Univ. of North Carolina Press, 1944.

Brems, Hans. *Product Equilibrium under Monopolistic Competition.* Cambridge, Harvard Univ. Press, 1951.

Burns, Arthur F. "Economic Research and the Keynesian Thinking of Our Times," *Twenty-sixth Annual Report of the National Bureau of Economic Research.* New York, 1946.

Calahan, Harold A. *Rigging.* New York, Macmillan, 1940.

Chamberlin, Edward H. "The Product as an Economic Variable," *Quarterly Journal of Economics,* LXVII, Feb., 1953.

——— "Product Heterogeneity and Public Policy," *American Economic Review: Papers and Proceedings,* XL, May, 1950.

——— *The Theory of Monopolistic Competition.* 6th ed. Cambridge, Harvard Univ. Press, 1948.

Clark, John Bates. *The Distribution of Wealth.* New York, Macmillan, 1900.

Clark, John Maurice. *Alternative to Serfdom.* New York, Knopf, 1948.

——— *Preface to Social Economics,* New York, Farrar & Rinehart, 1936.

——— "Realism and Relevance in the Theory of Demand," *Journal of Political Economy,* LIV, Aug., 1946.

——— *Studies in the Economics of Overhead Costs.* Chicago, Univ. of Chicago Press, 1923.

——— "Toward a Concept of Workable Competition," *American Economic Review,* XXX, June, 1940.

Cooley, Charles Horton. *Social Process.* New York, Scribner, 1918.

——— *Sociological Theory and Social Research.* New York, Henry Holt, 1930.

Cournot, Augustin. *Researches into the Mathematical Principles of the Theory of Wealth*. New York, Macmillan, 1897.
Davis, Kingsley. *Human Society*. New York, Macmillan, 1949.
Dewey, John. *How We Think*. New York, Heath, 1933.
——— *Human Nature and Conduct*. New York, Henry Holt, 1922.
Duesenberry, James S. *Income, Saving, and the Theory of Consumer Behavior*. Cambridge, Harvard Univ. Press, 1949.
Edwards, Corwin D. "Can the Antitrust Laws Preserve Competition?" *American Economic Review: Papers and Proceedings*, XXX, March, 1940.
——— *Maintaining Competition*. New York, McGraw-Hill, 1949.
Ely, Richard T., and others. *Outlines of Economics*. 5th ed. New York, Macmillan, 1930.
Enke, Stephen. "Profit Maximization under Monopolistic Competition," *American Economic Review*, XXXI, June, 1941.
Frank, Lawrence K. *Society as the Patient*. New Brunswick, Rutgers Univ. Press, 1948.
Galbraith, J. K. "Monopoly Power and Price Rigidities," *Quarterly Journal of Economics*, L, May, 1936.
Gambs, John S. *Beyond Supply and Demand*. New York, Columbia Univ. Press, 1946.
Georgescu-Roegen, N. "The Pure Theory of Consumer's Behavior," *Quarterly Journal of Economics*, L, Aug., 1936.
Gordon, Robert A. *Business Leadership in the Large Corporation*. Washington, Brookings Institution, 1945.
Graham, Frank D. *Social Goals and Economic Institutions*. Princeton, Princeton Univ. Press, 1942.
Griffin, Clare E. *Enterprise in a Free Society*. Chicago, Irwin, 1949.
Hayek, Friedrich A. *Individualism and Economic Order*. Chicago, Univ. of Chicago Press, 1948.
Hayes, Samuel P., Jr. "Potash Prices and Competition," *Quarterly Journal of Economics*, LVII, Nov., 1942.
Hotelling, Harold. "Stability in Competition," *Economic Journal*, XXXIX, March, 1929.
Jevons, W. Stanley. *The Theory of Political Economy*. 4th ed. London, Macmillan, 1911.
Kaldor, Nicholas. "Professor Chamberlin on Monopolistic and Imperfect Competition," *Quarterly Journal of Economics*, LII, May, 1938.
Klineberg, Otto. *Social Psychology*. New York, Henry Holt, 1940.
Knight, Frank H. *The Ethics of Competition*. New York, Harper, 1935.
——— "Realism and Relevance in the Theory of Demand," *Journal of Political Economy*, LII, Dec., 1944.

——— *Risk, Uncertainty, and Profit.* Boston, Houghton Mifflin, 1921. Reprinted by the London School of Economics, London, 1933.

Lange, Oskar, and Fred M. Taylor. *On the Economic Theory of Socialism.* Minneapolis, Univ. of Minnesota Press, 1938.

Machlup, Fritz. "Monopoly and Competition; a Classification of Market Positions," *American Economic Review,* XXVII, Sept., 1937.

Malinowski, Bronislaw. "The Group and the Individual in Functional Analysis," *American Journal of Sociology,* XLIV, May, 1939.

Marshall, Alfred. *Principles of Economics.* 8th ed. London, Macmillan, 1920.

Mason, Edward S. "Monopoly in Law and Economics," *Yale Law Journal,* Vol. 47, Nov., 1937.

——— "Various Views on the Monopoly Problem," *Review of Economics and Statistics,* XXXI, May, 1949.

Meade, J. E. *An Introduction to Economic Analysis and Policy.* 2d ed. London, Oxford Univ. Press, 1937.

Morgan, Theodore. "A Measure of Monopoly in Selling," *Quarterly Journal of Economics,* LX, May, 1946.

Mund, Vernon A. "Monopolistic Competition Theory and Public Price Policy," *American Economic Review,* XXXII, Dec., 1942.

Norris, Ruby Turner. *The Theory of Consumer's Demand.* 1st ed. New Haven, Yale Univ. Press, 1941.

Noyes, C. Reinold. *Economic Man in Relation to his Natural Environment.* 2 vols. New York, Columbia Univ. Press, 1943.

Parsons, Talcott. "Some Reflections on 'The Nature and Significance of Economics,' " *Quarterly Journal of Economics,* XLVIII, May, 1934.

Pigou, A. C. *The Economics of Welfare.* 4th ed. London, Macmillan, 1932.

Reder, Melvin Warren. *Studies in the Theory of Welfare Economics.* New York, Columbia Univ. Press, 1947.

Ricardo, David. *Principles of Political Economy and Taxation.* Edited by H. C. K. Gonner. London, G. Bell, 1932.

Robinson, Joan. *The Economics of Imperfect Competition.* London, Macmillan, 1933.

——— "What Is Perfect Competition?" *Quarterly Journal of Economics,* XLIX, Nov., 1934.

Samuelson, Paul A. *Foundations of Economic Analysis.* Cambridge, Harvard Univ. Press, 1947.

Schattschneider, E. E. *Party Government.* New York, Farrar & Rinehart, 1942.

Schumpeter, Joseph A. *Business Cycles.* 2 vols. New York, McGraw-Hill, 1939.

——— *Capitalism, Socialism, and Democracy.* New York, Harper, 1942.

Schumpeter, Joseph A. *The Theory of Economic Development.* Translated by Redvers Opie. Cambridge, Harvard Univ. Press, 1934.

Smith, Adam. *The Wealth of Nations.* New York, Modern Library, 1937.

Steiner, Peter O. "Program Patterns and Preferences, and the Workability of Competition in Radio Broadcasting," *Quarterly Journal of Economics,* LXVI, May, 1952.

Stigler, George J. *The Theory of Price.* 1st ed. New York, Macmillan, 1946.

Stolper, Wolfgang F. "The Possibility of Equilibrium under Monopolistic Competition," *Quarterly Journal of Economics,* LIV, May, 1940.

Sweezy, Paul M. "Expectations and the Scope of Economics," *Review of Economic Studies,* V, 1937–1938.

——— "Professor Schumpeter's Theory of Innovation," *Review of Economic Statistics,* XXV, Feb., 1943.

Triffin, Robert. *Monopolistic Competition and General Equilibrium Theory,* Cambridge, Harvard Univ. Press, 1940.

Wallace, Donald H. "Monopolistic Competition and Public Policy," *American Economic Review: Papers and Proceedings,* XXVI, March, 1936.

White, Horace G., Jr. "A Review of Monopolistic and Imperfect Competition Theories," *American Economic Review,* XXVI, Dec., 1936.

Index

Bei Fragen zur Produktsicherheit wenden Sie sich bitte an:
If you have any questions regarding product safety,
please contact:

Walter de Gruyter GmbH
Genthiner Straße 13
10785 Berlin
productsafety@degruyterbrill.com